THEORY OF FUNCTIONS

J. F. RITT
DAVIES PROFESSOR OF MATHEMATICS, COLUMBIA UNIVERSITY

REVISED EDITION

KING'S CROWN PRESS • NEW YORK

Copyright 1947 by

Joseph F. Ritt

FIRST EDITION 1946
First printing 1946

REVISED EDITION 1947
First printing 1947
Second printing 1949

KING'S CROWN PRESS
is a subsidiary imprint of Columbia University Press established for the purpose of making certain scholarly material available at minimum cost. Toward that end, the publishers have adopted every reasonable economy except such as would interfere with a legible format. The work is presented substantially as submitted by the author, without the usual editorial and typographical attention of Columbia University Press.

Composed on the Vari-Typer by Marie Russell
Diagrams prepared by Angela Pelliciari

Published in Great Britain and India
by Geoffrey Cumberlege, Oxford University Press
London and Bombay

Manufactured in the United States of America

Preface

There is presented herewith the basic material of a course in the theory of functions which I have given several times at Columbia University during the past twenty years. This course, which is given annually, is a two-semester course with twenty eight periods of seventy five minutes in each semester. The emphasis is on the complex variable. From this standpoint, about twenty periods are given to the real variable and the remaining time to the complex.

By the end of the first semester, the student finds himself equipped for higher courses in the real variable, or for a course on existence theorems for differential equations. After the second semester, he can study almost any topic of mathematical analysis. Above all, he has learned what it means to understand mathematics and can deal with any mathematical book.

In treating the real number system, I have used the method of infinite decimals, rather than the Dedekind or Cantor theories. The decimals have the advantage of lacking profundity and of not putting the student through a mathematical revolution. The numbers stay quite what they always were, instead of becoming new and bizarre objects. The objection that the decimals employ the special radix ten does not strike me as important. In any case, after the student has seen how simple the matter really is, he can read the Dedekind and Cantor theories with very little effort.

There is hardly time in the course for a full treatment of topological questions. However, all geometrical questions are formulated in arithmetic terms and every topological assumption made is explicitly stated. Of course, the student quickly sees that topological considerations are important only for securing a rounded theory and may be disregarded as far as cases arising in the applications are concerned.

I wish to thank Mr. Milton Sobel for reading the proofs for the present edition.

New York, N. Y.
July, 1947

J. F. RITT

Contents

I. **The Real Number System** 1

   Introduction -- Counting -- Sets -- Sequences -- Digits -- Terminating Decimals -- Addition and Multiplication -- Infinite Decimals -- Non-Negative Numbers -- Bounds -- Upper Bounds -- A Fundamental Theorem -- Least Upper Bounds -- Round Numbers -- Addition of Non-Negative Numbers -- Multiplication of Non-Negative Numbers -- Subtraction -- Division -- Negative Numbers -- Rational Numbers -- Irrational Numbers -- Bounds -- The Least Upper Bound -- The Greatest Lower Bound.

II. **Theory of Limits** 11

   Absolute Value -- Convergent Sequences of Numbers -- Uniqueness of Limit -- The Fundamental Convergence Theorem (Cauchy) -- Limits of Sums, Products, Quotients.

III. **Linear Point Sets** 16

   Intervals -- Neighborhoods -- Limit Points -- The Theorem on Nested Intervals -- The Bolzano-Weierstrass Theorem -- Closed Sets -- Borel's Theorem -- Countability.

IV. **Functions and Continuity** 20

   Variable -- Function -- Continuity -- Dense Sets -- A Function Continuous at a Dense Set of Points and Discontinuous at a Similar Set -- Continuity of Sum and Difference of Two Functions -- Continuity of Product of Two Functions -- Continuity of Quotient -- Bounded Functions -- Boundedness of a Function Continuous on a Closed Interval -- Attainment of Bounds by a Function Continuous on a Closed Interval -- Attainment of All Values Intermediate between Two Values -- Uniform Continuity -- Monotonic Functions.

V. **The Derivative** 26

   Derivative -- Differentiability and Continuity -- The Derivative as a Function -- Derivatives at Extremities of Intervals -- Right-Hand and Left-Hand Derivatives -- Maxima and Minima -- Necessary Condition for a Maximum or a Minimum at an Interior Point of an Interval -- The Mean Value Theorem -- Functions with Zero Derivatives -- Increasing and Decreasing Functions.

VI. **Riemann Integration** 29

   The Integral -- Condition for Integrability -- Integrability in Subintervals -- Bounds for an Integral -- Continuity of Integral with Respect to Upper Limit -- Integrability of Sum of Two Functions -- Integrability of Continuous Functions -- Differentiability of the Integral of a Continuous Function -- Evaluation of Definite Integrals -- Composition of Intervals -- Discontinuous Functions.

VII. **Infinite Series of Numbers** 37

   Infinite Series -- Convergent Infinite Series -- A Necessary and Sufficient Condition for Convergence -- Addition of Convergent Series -- The nth Term -- The Remainder -- A Class of Convergent Series -- Series of Absolute Values -- Absolute Convergence -- Rearrangement of Terms of an Absolutely Convergent Series.

VIII. Sequences of Functions 42
    Convergent Sequences of Functions -- Uniform Convergence -- Example of a Uniformly Convergent Sequence Defined on an Interval -- Example of Non-Uniform Convergence -- Necessary and Sufficient Condition for the Uniform Convergence of a Sequence -- Continuity of Limit of a Sequence -- Example of a Sequence of Continuous Functions Whose Limit Is Not Continuous -- Integrability of Limits of Sequences -- Differentiability of Sequences of Functions.

IX. Infinite Series of Functions 48
    Infinite Series of Functions -- Uniform Convergence of Infinite Series -- The Weierstrass M-Test -- Continuity, Integrability, Differentiability -- A Function Discontinuous at a Dense Set of Points, but Integrable.

X. Functions of Two Variables 51
    Points -- Functions of Two Variables -- Rectangles -- Continuity -- Partial Derivatives -- The Complete Differential.

XI. Complex and Hypercomplex Numbers 55
    Complex Numbers -- Addition and Multiplication -- Subtraction and Division -- Change of Notation -- The Modulus -- Geometric Representation -- Modulus of Sum and Difference -- Hypercomplex Numbers and Linear Algebras.

XII. Limits and Point Sets (Complex Domain) 58
    Limits -- Operations with Sequences -- Point Sets in the Complex Domain -- The Theorem on Nested Rectangles -- Borel's Theorem.

XIII. Curves and Regions 62
    Interior, Exterior and Boundary Points -- Continuous Curves -- Open Regions -- Functions and Continuity: Definition of Function -- Continuity -- Uniform Continuity.

XIV. Derivatives 65
    Derivative -- Monogenicity -- The Cauchy-Riemann Equations -- The Laplace Equation -- Definition of Analytic Function -- Conformal Mapping.

XV. Continuous Curves 71
    Inverse Functions (Real Variable) -- On the Points of a Continuous Curve -- Equivalent Curves -- Simple Curves -- Representation of a Simple Curve.

XVI. Rectifiable Curves 75
    Segment and Length -- Inscribed Polygon -- Rectifiable Curves -- Condition for Rectifiability -- Functions of Bounded Variation -- Rectifiability and Functions of Bounded Variation.

XVII. Curvilinear Integrals 79
    Functions Continuous on a Curve -- Curvilinear Integral -- Condition for Integrability -- Existence Theorem -- An Example.

XVIII. Jordan Curves 83
    Definition -- Equivalent Jordan Curves -- Application to Integration.

XIX. Analysis Situs of the Triangle 86
    Statement of the Jordan Separation Theorem -- Straight Lines -- Triangles -- Interior and Exterior -- Decomposition into Four Triangles -- Integration.

XX. The Cauchy Integral Theorem for Triangles 91
An Inequality for Curvilinear Integrals -- Statement of the Cauchy Integral Theorem for the Case of a Triangle -- Two Curvilinear Integrals -- On Differentiable Functions -- A Theorem on Sequences of Triangles -- Proof of the Cauchy Integral Theorem for Triangles.

XXI. Extension of the Cauchy Integral Theorem to Polygons. 95
Simply-Connected Open Regions -- Polygons -- Statement of the Cauchy Integral Theorem for Polygons -- Replacement of the Polygon C by a Simpler Polygon -- Completion of Proof.

XXII. The Cauchy Integral Theorem for a Rectifiable Curve 99
Curves in Open Regions -- Open Regions, Curves and Approximating Polygons -- Approximation to Integrals -- The Cauchy Integral Theorem for a Rectifiable Curve.

XXIII. The Cauchy Integral Theorem for Several Contours 102
Sensed Circles -- Sense of a Jordan Curve -- Integrals along Sensed Curves -- Regions Bounded by Several Contours -- The Cauchy Integral Theorem for Several Contours.

XXIV. Preliminaries for Cauchy Integral Formula 104
Analyticity at a Point -- Combinations of Analytic Functions -- The Number π -- Integral of $1/(z-a)$.

XXV. The Cauchy Integral Formula and the Derivatives of an Analytic Function . . 108
Cauchy Integral Formula -- The First Derivative -- The Higher Derivatives of an Analytic Function -- Analytic Functions Defined by Contour Integrals.

XXVI. Infinite Sequences and Infinite Series of Analytic Functions 112
Uniform Convergence -- Continuity and Integrability -- Sequences of Analytic Functions -- The Higher Derivatives -- Series of Complex Numbers -- Series of Analytic Functions.

XXVII. Power Series. 116
Greatest Limit of a Sequence of Real Numbers -- Power Series -- Circle of Convergence.

XXVIII. Taylor's Expansion 119
Statement of Results -- Proof of Uniqueness -- A Preliminary -- Derivation of Expansion.

XXIX. Liouville's Theorem and the Fundamental Theorem of Algebra 122
Integral Functions and Liouville's Theorem -- The Fundamental Theorem of Algebra.

XXX. On the Zeros of Analytic Functions 124
The Chief Theorem -- Some Consequences of the Preceding Theorem.

XXXI. Laurent Series 126
Series of Negative Powers -- Statement of the Expansion Theorem of Laurent -- Proof of Uniqueness, Determination of the Coefficients -- Derivation of Expansion.

XXXII. Singularities of Analytic Functions 130
Functions Bounded in the Neighborhood of a Point -- Isolated Singularities -- Poles -- The Number Infinity -- Reciprocals of Functions -- Isolated Essential Singularities -- Behavior of a Function at ∞.

XXXIII. Products and Quotients of Analytic Functions 134
Multiplicity of a Value -- Product of Two Functions -- Quotient of Two Functions -- Principal Part of a Laurent Development.

XXXIV. Rational Functions 137
    Decomposition of a Rational Function into Partial Fractions -- Functions Whose Only Singularities Are Poles -- Frequency of Attainment of Values -- Polynomials -- Degree of a Rational Function.

XXXV. The Functions e^z, sin z, cos z 140
    The Definitions -- First Relations -- The Multiplication Formula for e^z -- The Relation $\sin^2 z + \cos^2 z = 1$ -- The Addition Theorems for sin z and cos z — Periodicity of sin z, cos z, e^z -- Other Relations -- Graphs -- Parametric Representation of Circle -- A Formula for the Length of a Rectifiable Curve -- Proof that $p = \pi$ -- Amplitudes of a Complex Number.

XXXVI. Periodic Functions 147
    Period Strip -- Fundamental Domain -- The Function e^z -- The Functions sin z and cos z

XXXVII. Indefinite Integrals, Logarithms. 150
    Independence of an Integral of Path of Integration -- The Indefinite Integral -- Logarithm of an Analytic Function -- Integral Functions Which Are Nowhere Zero -- The Development of Log $(1 + z)$.

XXXVIII. Infinite Products. 153
    Convergent Infinite Products -- Necessary and Sufficient Condition for Convergence -- Absolute Convergence -- Infinite Products of Analytic Functions.

XXXIX. The Weierstrass Factorization Theorem. 157
    Integral Functions with a Finite Number of Zeros -- On Integral Functions with an Infinite Number of Zeros -- Construction of a Function with Assigned Zeros -- Convergence Proof.

XL. Meromorphic Functions and Mittag-Leffler's Theorem 161
    Definition of Meromorphic Function -- Representation of a Meromorphic Function as a Quotient of Integral Functions -- Mittag-Leffler's Theorem.

XLI. Theory of Residues 164
    Residue at a Finite Point -- Residue at Infinity -- Formula for Residues within a Curve -- The Residues of a Rational Function -- On the Logarithmic Derivative of an Analytic Function -- The Fundamental Theorem of the Calculus of Residues -- Special Cases -- Application to the Fundamental Theorem of Algebra.

XLII. Certain Important Theorems 168
    Rouché's Theorem -- Application to the Fundamental Theorem of Algebra -- Preservation of Neighborhoods -- On the Maximum of the Modulus of an Analytic Function -- On the Minimum of the Modulus of an Analytic Function -- Fundamental Theorem of Algebra -- Sequence of Analytic Functions

XLIII. Variation of the Amplitude of a Continuous Function along a Continuous Curve . 171
    Variation of Amplitude -- On the Zeros of an Analytic Function.

XLIV. The Functions $\sqrt[n]{z}$, log z 174
    The Function $\sqrt{z}$ -- The Function log z.

XLV. Analytic Continuation 177
    Elements and Their Continuations -- Monogenic Analytic Function -- Continuation along a Curve -- Singular Points -- Singular Points on a Circle of Convergence -- Natural Boundaries.

I

The Real Number System

INTRODUCTION

1. The theory of functions can be approached only on the basis of accurate notions concerning the real number system. In any treatment of the number system, there will be ideas which have to be taken for granted. The choice of the ideas which are to be considered as primitive depends always on questions of expediency. In what follows, we shall explain what we understand to be known in advance and we shall add a set of definitions and proofs which will furnish us with an adequate theory of the real number system.

COUNTING

We accept as a primitive idea the idea of counting. That is, we assume familiarity with the cardinal numbers

$$1, 2, 3, \ldots, n, \ldots$$

to which we attribute intuitive meaning and which we shall use to measure the multiplicities of collections of objects.

SETS

3. A set of objects will be called a *finite* set if there exists a cardinal number n such that the set contains exactly n objects.

A set of objects will be called an *infinite* set if, given any cardinal number n, the set contains more than n objects. For instance, the set of cardinal numbers is an infinite set.

SEQUENCES

4. Let n be any cardinal number. By a *sequence of n objects*, we shall mean a set of n objects.

$$O_1, O_2, \ldots, O_n,$$

arranged in a one-to-one correspondence with the first n cardinal numbers. Such a sequence will be said to be finite.

By an *infinite sequence* of objects, we shall mean an infinite set

$$O_0, O_2, \ldots, O_n, \ldots,$$

which *can be* arranged, and *is* arranged, in a one-to-one correspondence with the complete set of cardinal numbers.

DIGITS

5. Independently of their appearance in the symbols for the cardinal numbers, we shall employ the *digits* 0, 1, 2, 3, 4, 5, 6, 7, 8, 9 as *pure symbols*, with which we shall play a *game* which will be called *arithmetic*. The game will consist in performing certain operations with entities, called *numbers*, which will be constructed with the digits.

TERMINATING DECIMALS

6. By a *terminating decimal* we shall mean a finite sequence of digits with a decimal point written after some one of them, the first digit of the sequence being distinct from zero if the decimal point does not follow it immediately.

Examples: 56.47; 0.3420; 0.00.

If no digit follows the decimal point, the decimal point may be omitted with no danger of confusion. Thus, we may write 65 instead of 65..

For the present, we shall say "decimal" rather than "terminating decimal."

We shall regard the decimals as pure symbols, possessing no quantitative significance.

We shall now explain the meaning of the word *equal* as applied to two decimals. Let a and b be two decimals. If b is either identical with a or obtained from a by adjoining a set of zero digits to a, a and b will be called *equal*. Thus 5.4 and 5.400 are equal. It is to be emphasized that the adjective *equal* is merely a convenient word for describing the relationship just set forth. No meaning other than that given is to be read into our language. The word "equal," whose meaning is thus purely technical, will justify itself in later developments.

Two decimals which are not equal will be called *unequal*.

The terms "greater than" and "less than" as applied to two unequal decimals, will be employed in the manner which one would expect and we shall not use the space or time to give formal definitions for them. Thus,

$$64 > 53.1 \qquad 0.32 < 0.323.$$

The student already knows how to compare two unequal decimals and all that is important is to consider the comparison as a type of game playing rather than as an operation with concrete quantitative significance.

ADDITION AND MULTIPLICATION

7. We perform the operations of addition and multiplication upon decimals exactly as in popular arithmetic, with the distinction that we regard the operations as moves in a game.

Addition has the properties:
1. $a + b = b + a$ (Commutativity)
2. $a + (b + c) = (a + b) + c$ (Associativity)

Multiplication has the properties:
1. $ab = ba$ (Commutativity)
2. $a(bc) = (ab)c$ (Associativity)
3. $a(b + c) = ab + ac$ (Distributivity)

We note that, if $a = b$, $c = d$, then $a + c = b + d$, $ac = bd$.

INFINITE DECIMALS

8. Consider a symbol composed of an infinite sequence of digits with a decimal point written after some one digit, the first digit being distinct from zero if the decimal point does not follow it immediately. If there is no digit in the sequence such that every digit which follows is the digit 9, the symbol will be called an *infinite decimal*.

Examples: 2.333...
0.1212...
0.0000...

The symbol 2.43999...9... is not an infinite decimal.

In a decimal such as 64.24000..., which concludes with zero digits exclusively, the zero digits with which the decimal concludes may be omitted with no danger of confusion. In particular, we shall write 0 for 0.00....

When we say that two infinite decimals are "equal," we mean that they are identical. Two (infinite) decimals which are not identical will be called *unequal*. For two unequal decimals, we use the terms "greater than" and "less than" in the manner which the reader would propose himself if asked to establish meanings.

If $a > b$ and $b > c$, then $a > c$.

NON-NEGATIVE NUMBERS

9. An infinite decimal will be called a *non-negative number*. A non-negative number distinct from 0 will be called *positive*.

BOUNDS

10. Let E be any finite or infinite set of non-negative numbers. If there exists a positive number G such that every number of E is less than G, we shall call E *bounded*.

Thus, the set of non-negative numbers which do not exceed 3 (that is, 3.00...) is bounded. The sequence

$$2, 2.1, 2.11, 2.111, \ldots$$

is bounded. The sequence

$$1, 2, 3, \ldots, n, \ldots$$

is not bounded.

Every finite set of non-negative numbers is bounded.

UPPER BOUNDS

11. Let E be any bounded set of non-negative numbers. There exist non-negative numbers G such that, for every number a in E, one has

$$a \leq G.$$

Every such number G will be called an *upper bound* of E. If G is an upper bound of E and if $G' > G$, then G' is also an upper bound of E.

A FUNDAMENTAL THEOREM

12. *Theorem: Let E be a bounded set of non-negative numbers. Then there exists an upper bound of E which is less than every other upper bound of E.*

Remark: This is noteworthy because there exist sets of non-negative numbers which contain no least number. For instance, the sequence

$$1, 0.1, 0.01, 0.001, \ldots$$

has no least number.

Proof: By the *integral part* of a non-negative number a, we shall mean the number secured by replacing every digit which follows the decimal point by zero.

Considering the bounded set E, we examine the integral parts of the numbers of E. The boundedness of E certainly implies that there are only a finite number of distinct integral parts. We consider those numbers of E which have the greatest integral parts of all of the numbers of E. These numbers constitute a set which we shall E_0. The numbers of E_0 all have the same integral part. Let α represent the finite sequence of digits which precedes the decimal in that unique integral part.

From the numbers of E_0, let us select those for which the digit in the first decimal place is as great as it can be. Let β_1 be that greatest digit. The numbers of E_0 in which β_1 appears in the first decimal place constitute a set E_1.

From the numbers in E_1, we select those whose digit in the second decimal place is a maximum, say β_2. We continue in this fashion, selecting an infinite sequence of digits:

$$\beta_1, \beta_2, \ldots, \beta_n, \ldots$$

Consider now the symbol

(1) $\qquad \alpha . \beta_1 \beta_2 \beta_3 \ldots$

in which a decimal point follows α. Suppose first that there is no digit β_1 in the symbol such that $\beta_j = 9$ for $j > i$. Then the symbol (1) is a non-negative number. We designate this number by M. Very plainly, no number in E exceeds M. Thus, M is an upper bound for E. We have to show that M is less than any other upper bound. This is certainly true if M is zero. Suppose that M is positive. We shall show that no number less than M is an upper bound. Let M_1 be some number less than M. Then there must be some number

$$\alpha \cdot \beta_1 \beta_2 \ldots \beta_r 000\ldots$$

which exceeds M_1. On the other hand, the set E_r is made up of numbers no less than $\alpha \cdot \beta_1 \beta_2 \ldots \beta_r 000\ldots$. Thus M_1 is not an upper bound for E.

Suppose now that there is a digit β_i in (1) such that $\beta_j = 9$ for $j > i$. We consider the terminating decimal $\alpha \cdot \beta_1 \ldots \beta_i$ and add to it $0.00\ldots01$, where the digit 1 stands in the ith decimal place. The sum is a terminating decimal, which, if an infinite sequence of digits 0 is adjoined to it, produces a non-negative number. This non-negative number, we designate by M. The number M is seen as above, to be an upper bound of E which is less than any other upper bound.

LEAST UPPER BOUNDS

13. Let E be a bounded set of non-negative numbers. The number M whose existence was proved in §12 is called the *least upper bound* or *the upper bound* of E. Evidently, if E contains positive numbers, M is positive. In that case M has the following properties, *which characterize it*:

a. For every number a of E, $a \le M$.

b. Given any number M_1 less than M, E contains a number greater than M_1.

ROUND NUMBERS

14. By a *round number*, we shall mean an infinite decimal like $34.72580000\ldots$, which concludes with zero digits exclusively.

We define addition for two round numbers as follows: Suppress the zero digits in both numbers from some point on. There result two terminating decimals. Add these, and complete the sum into a round number by adjoining zeros. The resulting round number will be called the *sum* of the two given round numbers. Multiplication is defined similarly. Evidently, addition and multiplication, for round numbers, have the properties stated in §7.

ADDITION OF NON-NEGATIVE NUMBERS

15. Let a and b be two positive numbers, which may or may not be round numbers.

Consider all round numbers less than a. They form a class which we shall call E_1. Similarly, let E_2 be the class of round numbers which are less than b.

We form now a class E, putting into E every round number which is the sum of a number in E_1 and a number in E_2. (See §14.) Let us prove that E is bounded. Let α be any round number which exceeds a, and β any round number which exceeds b. Then every number in E is less than $\alpha + \beta$. (Obvious on the basis of elementary arithmetic.) Thus E is bounded.

By §12, E has a *least upper bound* M. If a and b are both round numbers, M is seen easily to be the sum of a and b as defined in §14. When at least one of a and b is not round, M will be taken *by definition* as the sum of a and b.

16. By definition, if a is any non-negative number, we take

$$a + 0 = 0 + a = a.$$

17. Evidently, addition as defined in §§15, 16 is commutative. We shall prove it associative. Let a, b, c be three infinite decimals. We shall prove that

$$a + (b + c) = (a + b) + c.$$

The case in which one of the three numbers is zero is trivial. Accordingly, we assume a, b, c all positive.

Suppose that
$$a + (b + c) > (a + b) + c.$$
Let d be a round number such that
$$a + (b + c) > d > (a + b) + c.$$
Because $d < a + (b + c)$, there must be round numbers α and δ with
$$\alpha < a \qquad \delta < b + c,$$
such that
$$(1) \qquad d < \alpha + \delta.$$
Because $\delta < b + c$, there are round numbers β and γ with
$$\beta < b, \qquad \gamma < c,$$
such that
$$(2) \qquad \delta < \beta + \gamma.$$
By (1) and (2),
$$d < \alpha + (\beta + \gamma)$$
so that, since addition is associative for round numbers, we have
$$(3) \qquad d < (\alpha + \beta) + \gamma.$$
We shall prove that (3) is incompatible with the relation
$$d > (a + b) + c.$$
This will follow if we can show that $\alpha + \beta < a + b$. Let α_1 and β_1 be round numbers such that
$$\alpha < \alpha_1 < a, \qquad \beta < \beta_1 < b.$$
Then $\alpha + \beta < \alpha_1 + \beta_1$. By the definition of addition, $\alpha_1 + \beta_1 \leq a + b$. Thus $\alpha + \beta < a + b$.

Thus, we cannot have $a + (b + c) > (a + b) + c$. Similarly, one shows that the relation
$$a + (b + c) < (a + b) + c$$
is impossible. Hence,
$$a + (b + c) = (a + b) + c.$$

MULTIPLICATION OF NON-NEGATIVE NUMBERS

18. We first consider the multiplication of two positive numbers a and b.

Let E_1 and E_2 be defined as in §15. We form a class E, putting into E every round number which is the product of a number of E_1 by a number of E_2. Then E has a least upper bound M. If a and b are round numbers, M is ab defined as in §14. We define ab for all cases (round numbers or not) as M.

By definition, we take $a \cdot 0 = 0 \cdot a = 0$ for every non-negative a.

Multiplication as defined above is evidently commutative. The associativity is proved as in §17. We leave the proof of the distributivity law as an exercise.

Two Lemmas:

19. **Lemma 1.** *Let $c \geq a$, $d \geq b$, where a, b, c, d are non-negative numbers. Then $c + d \geq a + b$.*

Suppose first that a and b are positive. Then a + b is the least upper bound of a certain set of numbers E described as in §15. Plainly every number in E is contained in the set of which $c + d$ is the least upper bound. Then $c + d$ is an upper bound of E, so that $c + d \geq a + b$.

If a and b are both zero, the result is trivial. Let $a = 0$ and $b > 0$. It suffices then to observe that b is the least upper bound of all round numbers less than b and that every such round number is less than d. The remarks of the preceding paragraph then apply.

Lemma 2. Let a be non-negative and b positive. Then $a + b > a$.

Proof: We can certainly find a number d, less than b, of the form

$$0.00\ldots0100\ldots$$

such that $a + d > a$. By Lemma 1, $a + b \geq a + d > a$.

SUBTRACTION

20. Let a and b be positive numbers with $a > b$. We shall prove that *there exists one and only one positive number c such that $a = b + c$.*

Proof: It is easy to see that there exist positive round numbers γ such that $b + \gamma < a$. In short, there are such numbers γ of the type

$$0.00\ldots0100\ldots$$

By Lemma 2 of §19, we have $b + \gamma > \gamma$ for any γ. Hence every round number γ such that $b + \gamma < a$ is less than a. Thus, the set of round numbers γ such that $b + \gamma < a$ is a bounded set. Let c be the least upper bound of this set. Then $c > 0$. We shall prove that

$$a = b + c.$$

First we show that $a \geq b + c$. This we prove by considering any two round numbers β and γ which are less respectively than b and c and showing that $a > \beta + \gamma$. Because $\gamma < c$, there is a round number γ^1 greater than γ, such that $b + \gamma^1 < a$. Now $b + \gamma^1 \geq \beta + \gamma^1$ by Lemma 1 of §19. Thus $a > \beta + \gamma^1$. Again, $\beta + \gamma^1 > \beta + \gamma$, (round numbers), so that $a > \beta + \gamma$. Hence $a \geq b + c$.

Suppose now that $a > b + c$. We can certainly find a positive δ such that

$$a > (b + c) + \delta$$

There are such numbers δ of the type $0.00\ldots0100\ldots$. Then

(1) $\qquad a > b + (c + \delta)$

By Lemma 2 of §19, $c + \delta > c$. Let γ be any round number which is greater than c but less than $c + \delta$.

By (1) and Lemma 1 of §19, we have

$$a > b + \gamma.$$

On the other hand, the definition of c shows that, because $\gamma > c$, we have

$$b + \gamma \geq a.$$

Thus, we cannot have $a > b + c$. Then $a = b + c$.

It remains to prove that c is unique. Suppose that

$$a = b + c_1, \qquad a = b + c_2$$

with $c_2 > c_1$. By what precedes, there is a $d > 0$ such that $c_1 + d = c_2$. Then

$$a + d = (b + c_1) + d = b + (c_1 + d) = b + c_2 = a.$$

But Lemma 2 of §19 shows that $a + d > a$. Thus c is unique.

The quantity c determined above is called the *difference* between a and b and is denoted by $a - b$. We have seen that 0 is the only number which when added to a number a produces a. We define $a - a$, for any non-negative a, as 0. Similarly, because $a = 0 + a$, we write $a - 0 = a$.

DIVISION

21. Let a and b be two positive numbers. We wish to show that there is one and only one positive number c such that

$$a = bc.$$

There are positive round numbers γ such that $b\gamma < a$. It is easy to show that those round numbers are bounded. Let c be the least upper bound of the numbers γ. One proves, following the method of §20, that $a = bc$. The uniqueness is also proved as in §20. We call c the quotient of a by b and write

$$c = \frac{a}{b}.$$

For the case of $a = 0$, we write, for any $b > 0$,

$$\frac{0}{b} = 0.$$

NEGATIVE NUMBERS

22. Let

$$\alpha \cdot \beta_1 \beta_2 \ldots,$$

with α a finite sequence of digits, be any positive infinite decimal. Introducing a dash, (that is, ───), into our category of mathematical symbols, we form the symbol

$$\text{───} \alpha \cdot \beta_1 \beta_2 \ldots.$$

The symbol just written will be called a *negative number*.

We use the symbol -0 as an equivalent of 0. That is, the meaning of "-0" is to be 0.

The non-negative numbers and the negative numbers will be called *real numbers*.

The arithmetic operations will be performed upon real numbers in the manner taught in first courses in algebra. For instance, if a and b are positive numbers, we are to have, *by definition*,

$$a(-b) = (-a)b = -(ab); \quad (-a)(-b) = ab.$$

If $a > 0$, $b > a$, we are to have $a - b = -(b - a)$. One sees from the above examples how the arithmetic operations are to be performed in other cases.

It is seen without difficulty that the arithmetic operations on real numbers, described as above, have the commutative, associative and distributive properties mentioned in §7.

Let a be any negative number. Let $a = -b$, where b is a positive number. By $-a$, we shall mean b. Thus, by definition,

$$-(-2) = 2.$$

The real number system is ordered as follows. We say that $a > b$ if $a - b$ is positive. This is harmonious with the ordering which existed at the start among the non-negative numbers.

RATIONAL NUMBERS

23. A real number in which every digit which follows the decimal point is zero will be called an *integer*. A number which is equal to the quotient of two integers will be called a *rational number*. Thus,

$$\frac{4}{7}, \frac{-3}{2}, 0, 1$$

are rational numbers.

24. We shall show that *between any two numbers there lies a rational number*.

Let a and b be two numbers with $a > b$. First, let a and b be positive. Then, between a and b there lies a round number, and round numbers are rational. A little reflection will show that the result is true in all cases.

IRRATIONAL NUMBERS

25. There exist numbers which are not rational. An example will be given below. A number which is not rational will be called *irrational*.

Towards showing the existence of irrational numbers, we shall prove first that there exists a positive number whose square equals 2.

There are certainly positive numbers whose squares are less than 2. Let E be the class of all such numbers. Evidently E is bounded. Let a be the least upper bound of E. We shall prove that $a^2 = 2$.

First, let $a^2 > 2$. Let $b = \frac{a^2 - 2}{2a}$. Then $b > 0$. We have $b = \frac{a}{2} - \frac{1}{a}$ so that $b < \frac{a}{2} < a$. Hence, $a - b > 0$. Now

$$(a - b)^2 = a^2 - 2ab + b^2 = 2 + b^2 > 2.$$

Thus $a - b$, which is a positive number less than a, has a square which exceeds 2. This contradicts the fact that a is the least upper bound of E. (Note that, since $a - b < a$, E contains a number c which exceeds $a - b$. As c is in E, $c^2 < 2$. But, if $c > a - b$, we would have $c^2 > (a - b)^2 > 2$.)

Similarly, one cannot have $a^2 < 2$. Thus, $a^2 = 2$. We denote a by $\sqrt{2}$.

26. To prove that $\sqrt{2}$ is irrational, we assume that

(1) $$\sqrt{2} = \frac{p}{q}$$

with p and q integers. Evidently we may assume p and q both positive. We have

(2) $$\frac{p^2}{q^2} = 2,$$

so that

(3) $$p^2 = 2q^2.$$

Equation (3) asserts the existence of positive integers which have the property that their squares are double the squares of positive integers. The integer p, for instance, has this property. Of all positive integers having this property, let r be the least. Let

(4) $$r^2 = 2s^2$$

with s a positive integer. We see at once that r is even, for the square of an odd integer is odd. Let $r = 2t$ with t a positive integer. Then (4) gives

(5) $$4t^2 = 2s^2$$

or

(6) $$s^2 = 2t^2$$

This is absurd, because $s < r$. Thus, $\sqrt{2}$ is irrational.

27. We shall prove that *between any two numbers there lies an irrational number*. Let a and b be any two numbers with $a > b$. Let a_1 and b_1 be two rational numbers such that

$$b < b_1 < a_1 < a.$$

Now $b_1 < b_1 + \frac{a_1 - b_1}{\sqrt{2}} < b_1 + (a_1 - b_1) = a_1.$

(Note that $\sqrt{2} > 1$.) Because $a_1 - b_1$ is rational, $\frac{a_1 - b_1}{\sqrt{2}}$ is irrational. (Why?) Then $b_1 + \frac{a_1 - b_1}{\sqrt{2}}$ is irrational and lies between b and a.

BOUNDS

28. Let E be any set, finite or infinite, of numbers.

If there exists a number G, positive, negative or zero, such that every number of E is less than G, E will be said to be *bounded from above*.

If there exists a number G, positive, negative or zero, such that every number of E exceeds G, E will be said to be *bounded from below*.

A set which is bounded from above and bounded from below will be called *bounded*.

Examples: If E consists of the numbers x such that

$$-10 \leq x \leq 12,$$

E is bounded. The set E consisting of the two infinite sequences

$$1, 1.9, 1.99, 1.999, \ldots$$
$$-1, -1.9, -1.99, -1.999, \ldots$$

is bounded. Every finite set of numbers is bounded.

The sequence

$$5, 4, 3, 2, 1, 0, -1, -2, \ldots, -n, \ldots$$

is bounded from above but not from below. The set of all positive numbers is bounded from below but not from above. The totality of integers is not bounded in either direction.

THE LEAST UPPER BOUND

29. Let E be a set of numbers which is bounded from above. Every number G which has the property that for every number a of E we have

$$a \leq G$$

will be called an *upper bound of E*.

We shall prove the existence of a number M such that

(1) No number of E exceeds M.

(2) Given any number M_1 less than M, E contains a number which exceeds M_1.

Clearly, such a number M is an upper bound for E and is less than every other upper bound. We shall call M, when its existence is proved, the *least upper bound* or *the* upper bound of E.

Existence proof: First, suppose that E contains non-negative numbers and let E_1 be the set of non-negative numbers which are contained in E. Let M be the least upper bound of E_1 whose existence was proved in §12. Then M has the properties (1) and (2) above.

Now, suppose that E contains only negative numbers. Let -a, where a > 0, be some number of E. Let a be added to every number of E and let the resulting set of numbers be denoted by E^1. Let M^1 be the least upper bound of E^1, shown above to exist. Then $M = M^1 - a$ has the properties (1) and (2) relative to E.

Q.E.D.

THE GREATEST LOWER BOUND

30. Let E be a set of numbers which is bounded from below. Every number G which has the property that for every number a of E

$$a \geq G$$

will be called a lower bound of E.

We shall prove the existence of a number m such that:

(1) No number of E is less than m.

(2) Given any number m_1 greater than m, E contains a number which is less than m_1.

The number m, which is evidently a lower bound for E, and greater than any other such bound, will be called the *greatest lower bound* of E or *the* lower bound of E.

Existence proof: Let E^1 consist of the negatives of the numbers of E. Then E^1 is bounded from above. Let M^1 be the least upper bound of E^1. Then $m = -M^1$ has the properties (1) and (2) above.

Q.E.D.

II
Theory of Limits

ABSOLUTE VALUE

1. By the absolute value of a number a, we shall mean a if $a \geq 0$ and $-a$ if $a < 0$. The absolute value of a will be denoted by $|a|$.

Example: $|-2| = 2$, $|0| = 0$, $|\sqrt{2}| = \sqrt{2}$.

Theorem: $|a + b| \leq |a| + |b|$.

Proof: Clear.

Examples: $|3 + 2| = |3| + |2|$; $|3 + (-2)| < |3| + |-2|$; $|-3 + (-3)| = |-3| + |-3|$.

Theorem: $|a - b| \leq |a| + |b|$.

Proof: Clear.

Theorem: $|a - b| \geq |a| - |b|$.

Proof: $a = b + (a - b)$; $|a| \leq |b| + |a - b|$; $|a - b| \geq |a| - |b|$.

Theorem: $|ab| = |a| \, |b|$.

Theorem: Let $b \neq 0$. Then $\left|\dfrac{a}{b}\right| = \dfrac{|a|}{|b|}$.

CONVERGENT SEQUENCES OF NUMBERS

2. Consider an infinite sequence of numbers

(1) $\qquad\qquad a_1, a_2, \ldots, a_n, \ldots$

The sequence will be called *convergent* if there exists a number λ such that, for every $\varepsilon > 0$, a positive integer N exists such that, for every $n > N$,

$$|\lambda - a_n| < \varepsilon.$$

It is to be emphasized that N will as a rule have to be taken different for different numbers ε. The important thing is that when some ε is assigned, an N can be found for that ε.

When a number λ exists as above, it is called the *limit* of the sequence (1). The sequence is said to *have* or to *approach* or to *converge* to the limit λ.

Examples: $\dfrac{1}{2}, \dfrac{3}{4}, \dfrac{7}{8}, \ldots, 1-\dfrac{1}{2^n}, \ldots$ (Limit 1)

$\qquad\qquad 2, 2, 2, \ldots, 2, \ldots$ (Limit 2)

The sequences

$$1, 2, 3, \ldots, n, \ldots$$
$$1, -1, 1, -1, \ldots$$

do not converge.

A sequence which is not convergent is said to be *divergent*.

UNIQUENESS OF LIMIT

3. A sequence

$$(1) \qquad a_1, a_2, \ldots, a_n, \ldots$$

cannot have two distinct limits. Suppose, in fact, that λ and $\lambda' \neq \lambda$ are both limits of (1). For n large,

$$|\lambda - a_n| < \tfrac{1}{2} |\lambda' - \lambda|, \quad |\lambda' - a_n| < \tfrac{1}{2} |\lambda' - \lambda|.$$

Then

$$|\lambda' - \lambda| = |(\lambda' - a_n) - (\lambda - a_n)| \leq |\lambda' - a_n| + |\lambda - a_n|$$
$$< \tfrac{1}{2} |\lambda' - \lambda| + \tfrac{1}{2} |\lambda' - \lambda| = |\lambda' - \lambda|$$

which is absurd.

THE FUNDAMENTAL CONVERGENCE THEOREM (CAUCHY)

4. *Theorem: For the sequence*

$$(1) \qquad a_1, a_2, \ldots, a_n, \ldots$$

to converge, it is necessary and sufficient that for every $\varepsilon > 0$ a positive integer N exist such that, for every $n > N$ and for every positive integer p,

$$|a_{n+p} - a_n| < \varepsilon.$$

Proof:

(a) *Necessity.* Let the sequence have a limit λ. Let any $\varepsilon > 0$ be given. Let N be such that for $n > N$, $|\lambda - a_n| < \tfrac{\varepsilon}{2}$. Then

$$|a_{n+p} - a_n| = |(\lambda - a_n) - (\lambda - a_{n+p})|$$
$$\leq |\lambda - a_n| + |\lambda - a_{n+p}| < \tfrac{\varepsilon}{2} + \tfrac{\varepsilon}{2} = \varepsilon.$$

(b) *Sufficiency.* We shall say that a number G is greater than every a_n of sufficiently large n if an N exists such that $G > a_n$ for $n > N$.

We shall prove the existence of such numbers G for the sequence (1). This we accomplish by producing a number G which exceeds *every* a_n.

Let N be such that $|a_{n+p} - a_n| < 1$ for $n > N$ and p arbitrary. Then $|a_n - a_{N+1}| < 1$ for $n > N$, so that, for $n > N$,

$$|a_n| = |(a_n - a_{N+1}) + a_{N+1}| \leq |a_n - a_{N+1}| + |a_{N+1}| < 1 + |a_{N+1}|.$$

Hence, if G is greater than the greatest of the numbers

$$a_1, a_2, \ldots, a_N, 1 + |a_{N+1}|,$$

then G exceeds every a_n.

It is possible for a number G to exceed every a_n of sufficiently large n without exceeding every a_n.

In an entirely similar manner, we can find numbers G which are less than every a_n. (It suffices to take G less than the least of the numbers

$$a_1, a_2, \ldots, a_N, -(1 + |a_{N+1}|),$$

with N as above.) Hence there exist numbers which do not have the property of exceeding every a_n of sufficiently large n.

Let E be the class of numbers which have the property of exceeding every a_n of large n. A number which does not have this property is less than every number in E. Since there are numbers without the property, *E is bounded from below.*

Let λ be the greatest lower bound of E. We shall prove that λ *is a limit for the sequence (1)*. Consider any positive number ε.

Because $\lambda + \frac{\varepsilon}{2} > \lambda$, there are numbers in E less than $\lambda + \frac{\varepsilon}{2}$. Hence $\lambda + \frac{\varepsilon}{2}$ is itself in E, that is, at most a finite number of the a_n are greater than or equal to $\lambda + \frac{\varepsilon}{2}$.

Let N be such that for $n > N$, $a_n < \lambda + \frac{\varepsilon}{2}$.

Let M be such that $n > M$, $|a_{n+p} - a_n| < \frac{\varepsilon}{2}$.

Because $\lambda - \frac{\varepsilon}{4}$ is not in E, an infinite number of the a_n exceed or equal $\lambda - \frac{\varepsilon}{4}$.

Hence an infinite number of the a_n exceed $\lambda - \frac{\varepsilon}{2}$.

Let P, greater than N and M, be such that
$$a_P > \lambda - \frac{\varepsilon}{2}.$$

Then a_P lies between $\lambda - \frac{\varepsilon}{2}$ and $\lambda + \frac{\varepsilon}{2}$, so that
$$|a_P - \lambda| < \frac{\varepsilon}{2}.$$

For every n,
$$|a_n - \lambda| = |(a_n - a_P) + (a_P - \lambda)| \leq |a_n - a_P| + |a_P - \lambda| < |a_n - a_P| + \frac{\varepsilon}{2}.$$

Now $P > M$. Hence, if $n > P$
$$|a_n - a_P| < \frac{\varepsilon}{2}.$$

Thus, for $n > P$,
$$|a_n - \lambda| < \frac{\varepsilon}{2} + \frac{\varepsilon}{2} = \varepsilon,$$

so that λ is a limit of the sequence.

Q.E.D.

LIMITS OF SUMS, PRODUCTS, QUOTIENTS

5. *Theorem:* Let
$$a_1, a_2, \ldots, a_n, \ldots$$
$$b_1, b_2, \ldots, b_n, \ldots$$

converge respectively to λ_a and λ_b. Then
$$a_1 + b_1, a_2 + b_2, \ldots, a_n + b_n, \ldots$$
converges and its limit is $\lambda_a + \lambda_b$.

Proof: Let $\varepsilon > 0$ be assigned arbitrarily. Let N_a be such that
$$|\lambda_a - a_n| < \frac{\varepsilon}{2}$$
for $n > N_a$. Let N_b be such that
$$|\lambda_b - b_n| < \frac{\varepsilon}{2}$$
for $n > N_b$. Let N be the greater of N_a and N_b. Then, for $n > N$,
$$|\lambda_a + \lambda_b - (a_n + b_n)| = |(\lambda_a - a_n) + (\lambda_b - b_n)|$$
$$\leq |\lambda_a - a_n| + |\lambda_b - b_n| < \frac{\varepsilon}{2} + \frac{\varepsilon}{2} = \varepsilon.$$

Q.E.D.

Theorem: Hypothesis as in preceding theorem --

Then
$$a_1b_1, a_2b_2, \ldots, a_nb_n, \ldots$$
converges and its limits is $\lambda_a\lambda_b$.

Proof: Let G be any number greater than $|\lambda_a|$, $|\lambda_b|$ and unity. Let ε be any positive number less than unity. Let N be such that for $n > N$,
$$|\lambda_a - a_n| < \frac{\varepsilon}{3G}, \qquad |\lambda_b - b_n| < \frac{\varepsilon}{3G}.$$

Now
$$\lambda_a\lambda_b - a_nb_n = \lambda_a(\lambda_b - b_n) + \lambda_b(\lambda_a - a_n) - (\lambda_a - a_n)(\lambda_b - b_n),$$
so that, for $n > N$,
$$|\lambda_a\lambda_b - a_nb_n| \leq |\lambda_a||\lambda_b - b_n| + |\lambda_b||\lambda_a - a_n| + |\lambda_a - a_n||\lambda_b - b_n|$$
$$< G\frac{\varepsilon}{3G} + G\frac{\varepsilon}{3G} + \frac{\varepsilon}{3G} \cdot \frac{\varepsilon}{3G}$$
$$< \frac{\varepsilon}{3} + \frac{\varepsilon}{3} + \frac{\varepsilon}{3}.$$

(Note that $\frac{\varepsilon}{3G} \cdot \frac{\varepsilon}{3G} = \frac{\varepsilon}{3G^2} \cdot \frac{\varepsilon}{3} < \frac{1}{3} \; \frac{\varepsilon}{3} < \frac{\varepsilon}{3}$.)

Q.E.D.

Theorem: Hypothesis as in preceding theorems with restriction that no b_n is 0 and that $\lambda_b \neq 0$.

Then
$$\frac{a_1}{b_1}, \frac{a_2}{b_2}, \ldots, \frac{a_n}{b_n}, \ldots$$
converges and its limit is $\frac{\lambda_a}{\lambda_b}$.

Proof: Let $G > 1$ be such that $G > |\lambda_a|$, $G > |\lambda_b|$ and $\frac{1}{G} < |\lambda_b|$.

Let $0 < \varepsilon < 1$ and let N be such that for $n > N$,
$$|\lambda_a - a_n| < \frac{\varepsilon}{4G^3}, \qquad |\lambda_b - b_n| < \frac{\varepsilon}{4G^3}$$

Now
$$\frac{\lambda_a}{\lambda_b} - \frac{a_n}{b_n} = \frac{\lambda_b(\lambda_a - a_n) - \lambda_a(\lambda_b - b_n)}{\lambda_b b_n}$$
$$= \frac{\lambda_b(\lambda_a - a_n) - \lambda_a(\lambda_b - b_n)}{\lambda_b[\lambda_b - (\lambda_b - b_n)]}$$

Thus, for $n > N$,
$$\left|\frac{\lambda_a}{\lambda_b} - \frac{a_n}{b_n}\right| < \frac{G\frac{\varepsilon}{4G^3} + G\frac{\varepsilon}{4G^3}}{\frac{1}{G}[\frac{1}{G} - \frac{1}{2G}]}, \quad \text{(Note that } \frac{\varepsilon}{4G^3} < \frac{1}{2G}\text{)}$$
$$< \frac{\frac{\varepsilon}{2G^2}}{\frac{1}{2G^2}} = \varepsilon.$$

Q.E.D.

If $\lambda_b = 0$, various situations are possible.

Example 1:
$$\frac{1}{2}, \frac{1}{4}, \frac{1}{8}, \ldots, \frac{1}{2^n}, \ldots$$
$$-\frac{1}{2}, \frac{1}{4}, -\frac{1}{8}, \ldots, \frac{(-1)^n}{2^n}, \ldots$$

Here there is no limit for the sequence of quotients.

Example 2:
$$1, \frac{1}{2}, \frac{1}{3}, \ldots, \frac{1}{n}, \ldots$$
$$1, \frac{1}{2^2}, \frac{1}{3^2}, \ldots, \frac{1}{n^2}, \ldots$$

Again, no limit for the quotients.

Example 3:
$$1, \frac{1}{2}, \frac{1}{3}, \frac{1}{4}, \ldots$$
$$\frac{1}{2}, \frac{1}{4}, \frac{1}{6}, \frac{1}{8}, \ldots$$

There is a limit, which equals 2.

III

Linear Point Sets

INTERVALS

1. Let a, and b > a be two numbers. The set of numbers which are neither less than a nor greater than b will be called the *closed interval* (a, b). The set of all numbers which exceed a and are less than b will be called the *open interval* (a, b). By the *length* of (a, b), (open or closed), we mean the number b - a.

NEIGHBORHOODS

2. In what follows, the word *point* will be used as an equivalent for *real number*.

Consider a point $\underline{a}$. Let ε be any positive number. The set of points in the open interval $(a - \varepsilon, a + \varepsilon)$ is called a *neighborhood of a*. Thus, every point has an infinite number of neighborhoods, one for each ε.

LIMIT POINTS

3. Let E be any set of points. A point a, not necessarily belonging to E, is called a *limit point* of E if every neighborhood of a contains a point of E distinct from a. If a is a limit point of E, every neighborhood of a will contain an infinite number of points of E.

Examples: Every point is a limit point of the set of all real numbers or of the set of all rational numbers or of the set of all irrational numbers. A finite point set has no limit point. The point 0 is a limit point for the set

$$1, \frac{1}{2}, \frac{1}{4}, \frac{1}{8}, \ldots$$

THE THEOREM ON NESTED INTERVALS

4. A closed interval (a, b) will be said to be contained in a given closed interval (c, d) if every point of (a, b) is a point of (c, d). The points a and c may coincide, and so may b and d. By a *point contained in the closed interval* (a, b), we mean any point of the closed interval. Thus, even a and b are contained in (a, b).

Theorem: Let

$$(a_1, b_1), (a_2, b_2), \ldots, (a_n, b_n), \ldots$$

be an infinite sequence of closed intervals, each one after the first being contained in the preceding one, whose lengths approach zero. There exists one and only one point which is contained in every interval of the sequence.

Proof: Consider the sequence of points

$$(1) \qquad a_1, a_2, \ldots, a_n, \ldots,$$

that is, the left ends of the intervals. This is a convergent sequence, since a_n and all succeeding points of (1) lie in (a_n, b_n). (Cauchy's convergence theorem.) Let P be the limit of the sequence (1). Then P is contained in every interval (a_n, b_n). For, let us imagine that P is not contained in (a_j, b_j). Then a_j and the points which follow it in (1) cannot form a sequence converging to P. This proves that P lies in every interval.

Furthermore, if Q is a point distinct from P, then, if the length of (a_n, b_n) is less than $|Q - P|$, as it will be when n is large, Q cannot lie in (a_n, b_n). Hence P is the only point contained in all of the intervals.

THE BOLZANO-WEIERSTRASS THEOREM

5. In harmony with what has been said concerning the meaning of the word *point*, a *bounded set of points* will mean a bounded set of numbers.

Theorem: Every bounded infinite set of points has at least one limit point.

Proof: Let E be an infinite set of points contained in the closed interval (a, b). (Note that every bounded set is contained in some closed interval.) We represent (a, b) by I_1. Let c be the midpoint of (a, b). Then either the closed interval (a, c) contains an infinite number of points of E or the closed interval (c, b) does. Pick out that one of the two intervals which contains an infinite number of points of E, or either, if both contain an infinite number of points of E. Let the interval selected be represented by I_2. The length of I_2 is half that of I_1. Similarly, there is an interval I_3 whose length is half that of I_2 which contains an infinite number of points of E. We form in this way an infinite sequence of intervals

$$I_1, I_2, \ldots, I_n, \ldots$$

each contained in, and each half the length of, its predecessor. Each I_n contains an infinite number of points of E.

By §4, there is a point P which is contained in each of the intervals I_n. Evidently, P is a limit point of E.

Remarks: In the preceding proof, it may occur an infinite number of times that both half-intervals contain an infinite number of points of E. Unless some definite procedure is given for choosing one of the two intervals which are available, one will experience a sense of dissatisfaction as regards the actual existence of a set $I_1, I_2, \ldots$. This vagueness can be removed, for instance, by choosing the right half-interval when each half is available.

In some proofs involving an infinite number of selections, one is unable to furnish a definite procedure for the selections. In such cases one tolerates the resulting vagueness. This question is discussed in the theory of assemblages, where an assumption known as "Zermelo's axiom" is invoked to authorize the selections.

Q.E.D.

CLOSED SETS

6. A point set will be called closed if either

(a) *it has limit points and contains all of them*

or

(b) *it has no limit points.*

Examples of type (a):

(1) A closed interval.

(2) The set $0, 1, \frac{1}{2}, \frac{1}{4}, \frac{1}{8}, \ldots$

(3) The set $\frac{1}{2}, \frac{3}{4}, \frac{7}{8}, \ldots$

$$1$$
$$0$$

$$\frac{1}{4}, \frac{1}{8}, \frac{1}{16}, \ldots$$

Examples of sets which are not closed:

(1) All rational numbers.

(2) An open interval.

(3) The point set $\frac{1}{2}, \frac{1}{4}, \frac{1}{8}, \ldots$

BOREL'S THEOREM

7. Let (a, b) be an interval, *open or closed*. A point c will be said to be *interior* to (a, b) if

$$a < c < b.$$

Borel's Theorem: Let E be a closed and bounded point set. Let Σ be an infinite set of intervals (open or closed) such that each point of E is interior to at least one of the intervals of Σ. Then there exists a finite subset of the intervals of Σ such that each point of E is interior to at least one interval of the finite subset.

Proof: We suppose that we have an E and a Σ described as in the hypothesis, and that the conclusion does not hold for this E and Σ. We shall produce a contradiction.

Let (a, b) be a closed interval containing E. (Note that E is bounded.) We denote (a, b) by I_1. Let c be the midpoint of (a, b). Consider the closed intervals (a, c) and (c, b). There must be one of these intervals such that the points of E lying in it cannot be covered by a finite number of intervals chosen from Σ. (When we say that a set of intervals "covers" a set of points, we mean that each point of the point set is interior to at least one of the given intervals.)

Let I_2 be that one of the two intervals (a, c) and b, c) whose points of E cannot be so covered, or either, if the covering is impossible for both intervals. We find similarly a closed interval I_3, which is half of I_2, such that the points of E lying in I_3 cannot be covered by a finite number of the intervals of Σ. We build in this way an infinite sequence

$$I_1, I_2, \ldots, I_n, \ldots$$

By §4, there is a point P which is contained in all of the intervals I_n.

Of course, P is a limit point of E, for there is an interval I_n in every neighborhood of P and every I_n contains an infinite number of points of E. Since E is closed, P belongs to E.

There is an interval in Σ which contains P in its interior. This interval contains every I_n of sufficiently large n in its interior, since the length of I_n approaches zero as n increases. We have thus a contradiction of the inference that the points of E lying in I_n cannot be covered by a finite number of intervals taken from Σ. hence a contradiction of the assumption that E cannot be so covered.

Q.E.D.

COUNTABILITY

8. An infinite set is said to be *countable* if it possible to establish a one-to-one correspondence between its elements and the positive integers; that is, if the elements of the set can be arranged in a sequence. A set which is not countable is called *uncountable*.

Theorem: The positive rational numbers are countable.

Proof: We arrange the rational numbers $\frac{p}{q}$ with $p > 0$, $q > 0$ (p and q integers), according to the size of p + q, taking care not to repeat any rational number. We secure thus a sequence:

$$\frac{1}{1}; \frac{1}{2}, \frac{2}{1}; \frac{1}{3}, \frac{3}{1}; \frac{1}{4}, \frac{2}{3}, \frac{3}{2}, \frac{4}{1}; \ldots$$

It is clear that all positive rational numbers will be included in this sequence.

Theorem: The real numbers are uncountable.

It will suffice to show that the numbers in the closed interval (0, 1) are uncountable, for an infinite subset of a countable set is evidently countable.

We shall prove that, given any infinite sequence of numbers

(1) $$a_1, a_2, \ldots, a_n, \ldots$$

in (0,1), there is a number in (0,1) which is not in the sequence (1). Let a_n have the decimal representation

$$0 \cdot \beta_{1n} \beta_{2n} \beta_{3n} \ldots$$

Let α_1 be any digit which is distinct from β_{11} and from 9. Let α_2 be any digit distinct from β_{22} and from 9. In general, let α_i be a digit distinct from β_{ii} and from 9. Then the number

$$0 \cdot \alpha_1 \alpha_2 \ldots \alpha_i$$

which lies in (0,1) is distinct from every a_n of (1).

IV

Functions and Continuity

VARIABLE

1. We consider any set of numbers E. We introduce a symbol x, which we use to represent a point of E when such a point is called to our attention. The symbol x will be called a *variable*. The set is called the *domain* of x. For domains, we shall, as a rule, use intervals.

FUNCTION

2. Suppose that with every point x of E there is associated a number whose value is known as soon as x is known. We introduce a symbol y and use it to represent the number associated with x. We call the symbol y a *function* of the symbol x. Sometimes, we write $y = f(x)$. The function y is said *to be defined on the domain* E.

CONTINUITY

3. Let $y = f(x)$ be defined at every point of an open interval (a, b). Let x_0 be an *interior* point of (a, b). We shall say that $f(x)$ is *continuous* at x_0 if, for every $\varepsilon > 0$, a $\delta > 0$ exists such that, if $|h| < \delta$, we have

$$|f(x_0 + h) - f(x_0)| < \varepsilon.$$

Roughly speaking, when x is changed slightly from x_0, $f(x)$ changes slightly.

Suppose that $f(x)$ is defined at <u>a</u> as well as in the open interval. We shall say that $f(x)$ is continuous at <u>a</u> if, for every $\varepsilon > 0$, a $\delta > 0$ exists such that, if $0 < h < \delta$, $|f(a+h) - f(a)| < \varepsilon$. A similar definition is made for the point <u>b</u>.

If $f(x)$ is not continuous at x_0, $f(x)$ will be said to be *discontinuous* at x_0.

Examples of continuous functions: If $f(x)$ is a constant, it is continuous for every x. The same is true of $y = x$.

Examples of discontinuous functions:

(a) Let $y = 0$ when x is rational and $y = 1$ when x is irrational. Then y is discontinuous for every value of x.

(b) Let $y = x$ when $x \neq 0$. Let $y = 1$ for $x = 0$. Then y is continuous for every $x \neq 0$ and is discontinuous at $x = 0$.

DENSE SETS

4. A set of points is said to be *dense in an interval (a, b)* if every subinterval of (a, b) contains a point of the set. A set is called *every where dense among the real numbers*, or, more briefly, *every where dense*, or, still more briefly, *dense*, if every interval (a, b), arbitrarily small, contains a point of the set.

A FUNCTION CONTINUOUS AT A DENSE SET OF POINTS AND DISCONTINUOUS AT A SIMILAR SET

5. If x is positive and rational and equal to $\frac{p}{q}$, with p an integer, q a positive integer and $\frac{p}{q}$ in its lowest terms, we let $y = \frac{1}{q}$.

If x is positive and irrational, we let $y = 0$.

Clearly y is discontinuous for every rational value of x.

Let x be irrational. Let $\varepsilon > 0$ be given. Let n be an integer which exceeds $\frac{1}{\varepsilon}$. Then $\frac{1}{n} < \varepsilon$. Plot all rational numbers whose denominators are 1 or 2 or 3 or ... or n. There is an interval with x as midpoint which contains none of these rational numbers. Let δ be half the length of such an interval.

Suppose that $|h| < \delta$, then, if $x + h$ is rational, say $\frac{p}{q}$, we have $q > n$, so that

$$|f(x + h)| = \frac{1}{q} < \frac{1}{n} < \varepsilon.$$

Hence

$$|f(x + h) - f(x)| = |f(x + h) - 0| = |f(x + h)| < \varepsilon.$$

If $x + h$ is irrational, $f(x + h) - f(x) = 0 - 0 = 0$. Hence, for $|h| < \delta$, $|f(x + h) - f(x)| < \varepsilon$.

Q.E.D.

CONTINUITY OF SUM AND DIFFERENCE OF TWO FUNCTIONS

6. *Theorem:* If $f(x)$ and $g(x)$ are both continuous at a point x_0, then $f(x) + g(x)$ and $f(x) - g(x)$ are continuous at x_0.

Proof: We limit outselves to $f(x) + g(x)$. Let $\varepsilon > 0$ be assigned. Let $\delta_1 > 0$ be such that, for $|h| < \delta_1$, $|f(x_0 + h) - f(x_0)| < \frac{\varepsilon}{2}$, and let $\delta_2 > 0$ be such that, for $|h| < \delta_2$, $|g(x_0 + h) - g(x_0)| < \frac{\varepsilon}{2}$. Let δ be the lesser of δ_1 and δ_2 (either, if $\delta_1 = \delta_2$). Then, for $|h| < \delta$,

$$|[f(x_0 + h) + g(x_0 + h)] - [f(x_0) + g(x_0)]|$$
$$= |[f(x_0 + h) - f(x_0)] + [g(x_0 + h) - g(x_0)]|$$
$$< \frac{\varepsilon}{2} + \frac{\varepsilon}{2} = \varepsilon.$$

Q.E.D.

CONTINUITY OF PRODUCT OF TWO FUNCTIONS

7. *Theorem:* Let $f(x)$ and $g(x)$ be continuous at x_0. Then $f(x) g(x)$ is continuous at x_0.

Proof: Let $\varepsilon > 0$ be assigned. We assume, as we may, that $\varepsilon < 1$.

Let G be a number which exceeds $|f(x_0)|$, $|g(x_0)|$ and unity. Let $\delta > 0$ be such that, for $|h| < \delta$, $|f(x_0 + h) - f(x_0)| < \frac{\varepsilon}{3G}$, $|g(x_0 + h) - g(x_0)| < \frac{\varepsilon}{3G}$.

We have, for $|h| < \delta$,

$$|f(x_0 + h) g(x_0 + h) - f(x_0) g(x_0)| = |f(x_0) [g(x_0 + h) - g(x_0)]$$
$$+ g(x_0) [f(x_0 + h) - f(x_0)]$$
$$+ [f(x_0 + h) - f(x_0)] [g(x_0 + h) - g(x_0)]|$$
$$< G \cdot \frac{\varepsilon}{3G} + G \cdot \frac{\varepsilon}{3G} + \frac{\varepsilon}{3G} \cdot \frac{\varepsilon}{3G}$$
$$< \frac{\varepsilon}{3} + \frac{\varepsilon}{3} + \frac{\varepsilon}{3} = \varepsilon.$$

Q.E.D.

Theorem: Any polynomial $a_0 x^n + a_1 x^{n-1} + \ldots + a_n$ is continuous for every x.

This follows from the theorems on sums and products.

CONTINUITY OF QUOTIENT

8. *Theorem: Let $f(x)$ and $g(x)$ be continuous at x_0. Let $g(x_0) \neq 0$. Then $\frac{f(x)}{g(x)}$, which is defined on an interval containing x_0, is continuous at x_0.*

Proof left as exercise. See corresponding proof for limits.

BOUNDED FUNCTIONS

9. Let a function $f(x)$ be defined on a domain E. Let E' be the set of values which $f(x)$ assumes on E, that is, the set of values of $f(x)$ for the various values of x in E.

If E' is bounded from above, we call $f(x)$ *bounded from above on E*. We call the least upper bound of E' the *least upper bound of $f(x)$ on E*. We designate the least upper bound of $f(x)$ on E by M.

Examples:

(a) Let E be the closed interval (0, 1). Let $f(x) = x$. Then $M = 1$.

(b) Let E be the set of all positive numbers. Let $f(x) = \frac{1}{x}$. Then $f(x)$ is not bounded from above.

10. If E' is bounded from below, we call $f(x)$ *bounded from below on E*. We call the greatest lower bound of E' *the greatest lower bound of $f(x)$ on E*. We designate the greatest lower bound of $f(x)$ by m.

11. If E' is bounded, we call $f(x)$ *bounded on E*. Thus, $f(x)$ is bounded on E if and only if a $G > 0$ exists such that $|f(x)| < G$ for every x in E.

Examples of bounded functions:

(a) Let E be the set of all integers. Let $f(0) = 0$; $f(n) = 1 - \frac{1}{n}$ for $n > 0$; $f(n) = -1 - \frac{1}{n}$ for $n < 0$. Then

$$M = 1, \quad m = -1.$$

(b) Let E be the set $0 \leq x \leq 1$. Let $f(x) = x$. Then $M = 1, m = 0$.

(c) Let E be the set of all numbers. Let $f(x) = 1$. Then $M = m = 1$.

BOUNDEDNESS OF A FUNCTION CONTINUOUS ON A CLOSED INTERVAL

12. *Theorem: Let $f(x)$ be defined, and continuous, on a closed interval (a, b). Then $f(x)$ is bounded on (a, b).*

Note: when we say that $f(x)$ is continuous on (a, b), we mean that $f(x)$ is continuous at every point of (a, b).

Proof: Suppose that $f(x)$ is not bounded on (a, b). Let x_1 be a point of (a, b) such that $|f(x_1)| > 1$. Let x_2 be a point of (a, b) for which $|f(x_2)|$ exceeds $|f(x_1)|$ and 2. Let x_3 be a point of (a, b) such that $|f(x_3)|$ is greater than both $|f(x_2)|$ and 3. Continuing, we form an infinite sequence of distinct points,

$$(1) \qquad x_1, x_2, \ldots, x_n \ldots, \text{ with } |f(x_n)| > n.$$

By the Bolzano-Weierstrass theorem the points of the sequence (1) have at least one limit point. Let x_0 be such a limit point. Clearly, x_0 lies in (a, b).

Now $f(x)$ is continuous at x_0. Thus, we can find a $\delta > 0$ such that, if x' lies in (a, b) and $|x' - x_0| < \delta$, we have

$$|f(x') - f(x_0)| < 1.$$

Thus, for $|x' - x_0| < \delta$, we have

$$|f(x')| = |[f(x') - f(x_0)] + f(x_0)| < 1 + |f(x_0)|.$$

There are an infinite number of points x_n in the sequence (1) for which $|x_n - x_0| < \delta$. If we take a value of n greater than $1 + |f(x_0)|$ for which $|x_n - x_0| < \delta$, we have the contradiction that

$$|f(x_n)| > n > 1 + |f(x_0)|$$

Remarks: The preceding proof employs an infinite number of selections. One can give proofs which do not have this feature. One can, for instance, use the process of bisection used in proving the theorem on nested intervals.

Q.E.D.

ATTAINMENT OF BOUNDS BY A FUNCTION CONTINUOUS ON A CLOSED INTERVAL

13. *Theorem:* Let $f(x)$ be continuous on the closed interval (a, b). Let M and m be respectively the least upper and greatest lower bounds of $f(x)$ on (a, b). Then there is a value of x in (a, b) at which $f(x)$ equals M and a value of x in (a, b) at which $f(x) = m$.

Proof: We discuss M. The proof for m is similar.

Let us suppose that there is no value of x for which $f(x) = M$. By the nature of M, there is a point x_1 on (a, b) such that $f(x_1) > M - 1$. Naturally $f(x_1) < M$. Thus, there is a point x_2 on (a, b) such that $f(x_2)$ exceeds both $f(x_1)$ and $M - \frac{1}{2}$. Continuing, we form an infinite sequence of distinct points

(1) $x_1, x_2, \ldots, x_n, \ldots$ with $f(x_n) > M - \frac{1}{n}$.

Let x_0 be a limit point of the sequence (1). Of course, $f(x_0) < M$. We know that $f(x)$ is continuous at x_0. Let $\delta > 0$ be such that, for $|x' - x_0| < \delta$,

$$f(x') - f(x_0) < \frac{M - f(x_0)}{2}.$$

Then, for $|x' - x_0| < \delta$,

(2) $f(x') < f(x_0) + \frac{M - f(x_0)}{2} = M - \frac{M - f(x_0)}{2}$.

There are an infinite number of points x_n of the sequence (1) for which $|x_n - x_0| < \delta$. Let n be such that

$$n > \frac{2}{M - f(x_0)}$$

and such that $|x_n - x_0| < \delta$. Then

$$f(x_n) > M - \frac{1}{n} > M - \frac{M - f(x_0)}{2}.$$

This contradicts (2) above.

Q.E.D.

ATTAINMENT OF ALL VALUES INTERMEDIATE BETWEEN TWO VALUES

14. *Theorem:* Let $f(x)$ be continuous in the closed interval (a, b). Let $f(a) = A$, $f(b) = B$. Suppose that $A = B$. Let C be any number between A and B. Then there is a point c in (a, b) such that $f(c) = C$.

Proof: To fix our ideas, let $B > A$.

For x close to b, we have $f(x) > C$ since $C < B$. Let E be the class of all numbers ζ in (a, b) such that, for

$$\zeta \leq x \leq b$$

one has $f(x) > C$. Then E is bounded from below. Let c be the greatest lower bound of E. We say that $f(c) = C$.

Suppose first that $f(c) < C$. Then, for any x slightly greater than c, we have $f(x) < C$. It is easy to see, however, that any such x is in E, so that, for it, $f(x) > C$. Hence, we cannot have $f(c) < C$.

Suppose now that $f(c) > C$. Because $f(a) = A < C$, we have $c > a$. There is a $\delta > 0$ such that, in the closed interval $(c - \delta, c + \delta)$, one has $f(x) > C$. This means that $c - \delta$ is in E, which is absurd.

Q.E.D.

UNIFORM CONTINUITY

15. Let $f(x)$ be continuous in the closed interval (a, b). We shall show that, given any $\varepsilon > 0$, a $\delta > 0$ exists such that, for *any x at all in* (a, b) and for $|h| < \delta$, we have $|f(x + h) - f(x)| < \varepsilon$. We describe this situation by saying that $f(x)$ is *uniformly continuous* in (a, b). The concept of uniform continuity is useful in integration and in other problems.

Proof: Let $\varepsilon > 0$ be given. For every x in (a, b), there is a δ_x such that $|f(x + h) - f(x)| < \frac{\varepsilon}{2}$ for $|h| < \delta_x$. Put about each point x an interval with x for midpoint, of length δ_x. By Borel's Theorem (page 18), there exists a finite subset of the above intervals such that every point of (a, b) is interior to at least one interval of the finite subset. Let δ be any positive number less than half the length of the smallest interval of the finite subset. We say that δ answers to the assigned ε.

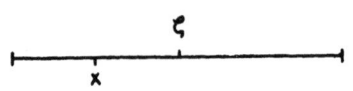

Let x be any point in (a, b). Consider one of the intervals of the finite set which contains x in its interior. (This interval is shown in the diagram.) Let ζ be the midpoint of this interval. Then

$$|x - \zeta| < \frac{1}{2} \delta_\zeta$$

Let h be such that $|h| < \delta$. Then

$$|(x + h) - \zeta| = |(x - \zeta) + h| \leq \frac{1}{2}\delta_\zeta + \delta < \frac{1}{2}\delta_\zeta + \frac{1}{2}\delta_\zeta = \delta_\zeta.$$

Hence

$$|f(x + h) - f(\zeta)| < \frac{\varepsilon}{2}.$$

Also, as $|(x - \zeta)| < \frac{1}{2}\delta_\zeta$,

$$|f(x) - f(\zeta)| < \frac{\varepsilon}{2}.$$

Then

$$|f(x + h) - (f(x)| = |[f(x + h) - f(\zeta)] - [f(x) - f(\zeta)]| < \frac{\varepsilon}{2} + \frac{\varepsilon}{2} = \varepsilon.$$

Q.E.D.

For a function continuous in an open interval, it may be impossible to find a single δ which answers to an assigned ε *for every x simultaneously*.

Example: Take $y = \frac{1}{x}$ with $0 < x \leq 1$.

MONOTONIC FUNCTIONS

16. Let $f(x)$ be defined in the interval (a, b), open or closed.

If, whenever $x_2 > x_1$, where x_1 and x_2 are in (a, b), we have

$$f(x_2) \geq f(x_1),$$

we say "$f(x)$ is non-decreasing in (a, b)."

If, whenever $x_2 > x_1$, we have
$$f(x_2) \leq f(x_1),$$
we call $f(x)$ *non-increasing* in (a, b).

A function which is non-decreasing or non-increasing in an interval is said to be *monotonic* in the interval.

If $f(x_2) > f(x_1)$ when $x_2 > x_1$, we call $f(x)$ *increasing*.

If $f(x_2) < f(x_1)$ when $x_2 > x_1$, we call $f(x)$ *decreasing*.

Theorem: Let $f(x)$ be continuous in the closed interval (a, b). Suppose that $f(x)$ assumes no value twice in (a, b). Then $f(x)$ is either increasing or decreasing in (a, b).

Remark: The assumption that (a, b) is closed is not essential and is made only for convenience.

Proof: Suppose that $f(b) > f(a)$. We shall prove that $f(x)$ is increasing.

Let $x_2 > x_1$. We cannot have $f(x_2) = f(x_1)$. We shall prove that $f(x_2) > f(x_1)$. Suppose that $f(x_2) < f(x_1)$. First, let us imagine that $f(x_2) < f(a)$. Then $f(x_2) < f(b)$. Between x_2 and b, $f(x)$ assumes all values between $f(x_2)$ and $f(b)$. In particular, it assumes the value $f(a)$. This contradicts the hypothesis.

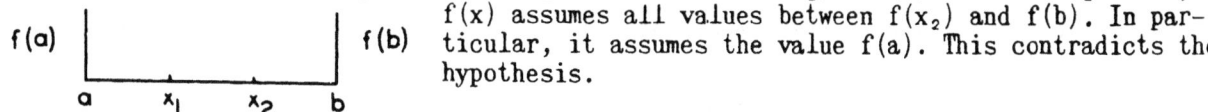

Suppose now that $f(x_2) > f(a)$. Between a and x_1, $f(x)$ assumes all values between $f(a)$ and $f(x_1)$, in particular, the value $f(x_2)$. This contradicts the hypothesis.

Q.E.D.

V

The Derivative

DERIVATIVE

1. Let $y = f(x)$ be a function whose domain of definition comprises intervals.

Let x_0 be a point throughout a neighborhood of which $f(x)$ is defined.

Consider the expression

$$\frac{f(x_0 + h) - f(x_0)}{h}$$

which, for every sufficiently small h distinct from zero, is a definite number. We call this expression a *difference quotient*.

The difference quotient is a function of h.

It may be that as h tends toward zero, the difference quotient tends toward a limit. That is, is, a number λ may exist such that, for every $\varepsilon > 0$, we can find a $\delta > 0$ such that, for $0 < |h| < \delta$,

$$\left|\frac{f(x_0 + h) - f(x_0)}{h} - \lambda\right| < \varepsilon.$$

Obviously at most one such number λ can exist. If such a λ exists, we call it the *derivative* of $f(x)$ at x_0. We say that $f(x)$ is *differentiable* at x_0.

DIFFERENTIABILITY AND CONTINUITY

2. *Theorem:* If $f(x)$ has a derivative at x_0, then $f(x)$ is continuous at x_0.

Proof: Let $\varepsilon > 0$ be given and let $\delta > 0$ be taken so that

$$\left|\frac{f(x_0) + h) - f(x_0)}{h} - \lambda\right| < \varepsilon$$

for $0 < |h| < \delta$. Then

$$\left|\frac{f(x_0 + h) - f(x_0)}{h}\right| < |\lambda| + \varepsilon$$

and

$$|f(x_0 + h) - f(x_0)| < |h| \, (|\lambda| + \varepsilon).$$

This last inequality proves the continuity of $f(x)$ at x_0.

The converse of the above theorem is not true. A function may be continuous at a point without having a derivative there.

THE DERIVATIVE AS A FUNCTION

3. Let $f(x)$ have a derivative at every point of an open interval (a, b). Then the derivative is a function of x defined on the interval. We designate the derivative by $f'(x)$ or by $\frac{dy}{dx}$.

The methods of the calculus for determining $\frac{dy}{dx}$ for the simple functions are entirely rigorous. We shall use freely the formulas:

$$\frac{d}{dx} x^n = nx^{n-1} \qquad \text{(n an integer)}.$$

$$\frac{d}{dx}(u+v) = \frac{du}{dx} + \frac{dv}{dx}$$

etc. The derivative of a constant is easily seen to be zero.

DERIVATIVES AT EXTREMITIES OF INTERVALS

4. Let $f(x)$ be defined on the closed interval (a, b). We consider

$$\frac{f(a+h) - f(a)}{h}$$

for positive values of h. If a number λ exists such that, for every $\varepsilon > 0$ a $\delta > 0$ can be found such that

$$\left| \frac{f(a+h) - f(a)}{h} - \lambda \right| < \varepsilon$$

for $0 < h < \delta$, we call λ the derivative of $f(x)$ at <u>a</u>. A similar definition is made for b.

RIGHT-HAND AND LEFT-HAND DERIVATIVES

5. Let $f(x)$ be defined for a neighborhood of x_0. It may be that $f(x)$ has no derivative at x_0 but that if <u>h</u> is allowed to approach zero through only positive values, the difference quotient approaches a limit. In that case, we say that the function has a *right-hand derivative* at x_0. Similarly, we define *left-hand derivative*.

Example: $y = x$ for $x \geq 0$ and $y = -x$ for $x < 0$. This function does not have a derivative for $x = 0$ but does have a right-hand and a left-hand derivative for $x = 0$.

MAXIMA AND MINIMA

6. Let $f(x)$ be defined on a closed interval (a, b). If the point x_0 in (a, b) is such that for some $\delta > 0$, $f(x_0) \geq f(x)$ if $|x - x_0| < \delta$, where it is understood that x is in (a, b), we say that $f(x)$ has a *maximum* at x_0.

Examples: (a) $y = 1 - x^2$; $x_0 = 0$.

(b) $y = x$ for interval $(0,1)$; $x_0 = 1$.

(c) $y = 1$; x_0 arbitrary.

The value of $f(x)$ at a point for which $f(x)$ has a maximum need not be the greatest value of $f(x)$ for the domain of definition.

Similarly, if a $\delta > 0$ exists such that $f(x_0) \leq f(x)$ if $|x - x_0| < \delta$, we say that $f(x)$ has a *minimum* at x_0.

NECESSARY CONDITION FOR A MAXIMUM OR A MINIMUM
AT AN INTERIOR POINT OF AN INTERVAL

7. *Theorem: Let $f(x)$ have a derivative at every point of a closed interval (a, b). Suppose that, at some point x_0 interior to (a, b), $f(x)$ has a maximum or a minimum. Then $f'(x_0) = 0$.*

Proof: To fix our ideas, let us suppose that $f(x)$ has a maximum at x_0. We have $f'(x_0) > 0$ or $f'(x_0) = 0$ or $f'(x_0) < 0$. Suppose that $f'(x_0) > 0$. Then, when h is very small,

$$\frac{f(x_0 + h) - f(x_0)}{h},$$

being very close to $f'(x_0)$, must be positive. Hence, if <u>h</u> is small and positive, $f(x_0 + h) - f(x_0)$ is positive. Thus, for <u>h</u> small and positive $f(x_0 + h) > f(x_0)$ so that $f(x)$ has no maximum at x_0.

Similarly, we cannot have $f'(x_0) < 0$.

Hence $f'(x_0) = 0$.

THE MEAN VALUE THEOREM

8. *Theorem: Let $f(x)$ be defined, and possess a derivative, throughout the closed interval (a, b). Then a point x_1, interior to (a, b), exists such that*

$$f(b) - f(a) = f'(x_1)(b-a).$$

Remark: It suffices actually for $f(x)$ to be continuous on the closed interval (a, b) and to have a derivative at all interior points.

Proof: Consider the function

$$\varphi(x) = f(x) - f(a) - \frac{x - a}{b - a}[f(b) - f(a)].$$

Evidently, $\varphi(x)$, like $f(x)$, is continuous throughout (a, b). We have $\varphi(a) = 0$, $\varphi(b) = 0$.

We shall prove that either $\varphi(x)$ has a maximum at some point interior to (a, b) or $\varphi(x)$ has a minimum at some interior point.

If $\varphi(x)$ is a constant, it will be zero throughout (a, b) so that there is both a maximum and a minimum at every interior point.

Suppose that $\varphi(x)$ is not a constant. Then either its least upper bound is positive or its greatest lower bound is negative. Suppose that the least upper bound is positive. By a previous theorem, there is a point x_0, necessarily interior to (a, b), at which $\varphi(x)$ attains its upper bound. Clearly, $\varphi(x)$ has a maximum at x_0. Similarly, if the greatest lower bound is negative, $\varphi(x)$ has a minimum at some interior point.

At an interior point which gives a maximum or a minimum, $\varphi'(x) = 0$.

Thus there is an x_1 such that $a < x_1 < b$ for which $\varphi'(x_1) = 0$. We have

$$\varphi'(x) = f'(x) - \frac{f(b) - f(a)}{b - a}$$

Hence

$$0 = f'(x_1) - \frac{f(b) - f(a)}{b - a}$$

which proves the mean value theorem.

FUNCTIONS WITH ZERO DERIVATIVES

9. *Theorem: If $f'(x)$ exists and is equal to zero throughout the closed interval (a, b), then $f(x)$ is constant on (a, b).*

Proof: Let x be such that $a < x \leq b$. Then x_1 exists, with $a < x_1 < x$, such that

$$f(x) - f(a) = f'(x_1)(x-a) = 0.$$

Hence $f(x) = f(a)$ throughout (a, b).

Remark: In the above theorem, as in the two which follow, the assumption that (a, b) is closed is convenient rather than necessary.

INCREASING AND DECREASING FUNCTIONS

10. *Theorem: Let $f'(x)$ exist and be positive throughout the closed interval (a, b). Then $f(x)$ is an increasing function on (a, b).*

Proof: As in §9.

Theorem: Let $f'(x)$ exist and be negative throughout the closed interval (a, b). Then $f(x)$ is a decreasing function on (a, b).

VI

Riemann Integration

THE INTEGRAL

1. We consider a function f(x), defined on a closed interval (a, b). Taking any positive integer n, we divide (a, b) into n equal or unequal parts by means of n + 1 points

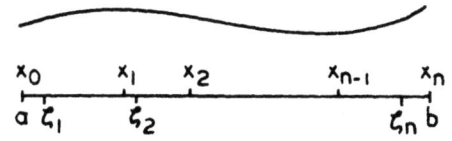

$$x_0 = a,\ x_1 > x_0, \ldots,\ x_{n-1} > x_{n-2},\ x_n = b > x_{n-1}.$$

In each interval (x_{i-1}, x_i), $i = 1, \ldots, n$, we take *arbitrarily* a point ζ_i. We consider the sum

(1) $\qquad f(\zeta_1)(x_1 - x_0) + f(\zeta_2)(x_2 - x_1) + \ldots + f(\zeta_n)(x_n - x_{n-1}).$

We write this sum more briefly

(2) $\qquad \sum_{i=1}^{n} f(\zeta_i)(x_i - x_{i-1}).$

By the *norm* of the sum (2), we shall mean the greatest of the quantities

$$x_i - x_{i-1},\ i = 1, \ldots, n.$$

We are going to introduce the notion of the approach to a limit of the sum (2) as n increases indefinitely and the lengths of the intervals (x_{i-1}, x_i) tend towards zero.

We shall say that the *sum (2) tends toward a limit as its norm tends towards zero* if a number I exists such that, for every $\varepsilon > 0$, a $\delta > 0$ exists such that, for

$$x_i - x_{i-1} < \delta,\ i = 1, \ldots, n,$$

we have, for any ζ's at all [ζ_i in (x_{i-1}, x_i)],

$$\left| I - \sum_{i=1}^{n} f(\zeta_i)(x_i - x_{i-1}) \right| < \varepsilon.$$

Obviously, at most one such number I can exist. If such an I exists, we shall say that f(x) is *integrable* in (a, b). We shall call I the *integral of f(x) from a to b*. We write

$$I = \int_a^b f(x)\ dx.$$

2. *Theorem:* If f(x) is integrable in (a, b), then f(x) is bounded in (a, b).

Proof: Suppose that a function f(x) is integrable in (a, b) but not bounded. Let $\int_a^b f(x)\ dx = I$.

An $\varepsilon > 0$ being given, let $\delta > 0$ be taken so that any sum (2) of norm less than δ differs from I by less than ε.

Take any subdivision into intervals

$$(x_0, x_1),\ (x_1, x_2), \ldots,\ (x_{n-1}, x_n)$$

with every $x_i - x_{i-1}$ less than δ.

Then there is one (at least) of the intervals (x_{i-1}, x_i) in which $f(x)$ is not bounded. To fix our ideas, suppose that $f(x)$ is not bounded in (x_0, x_1).

Let $\zeta_2, \zeta_3, \ldots, \zeta_n$ be chosen, and held fixed. Then $f(\zeta_2), \ldots, f(\zeta_n)$ are definite numbers.

Keep ζ_1 arbitrary for the present. Consider the sum

$$(3) \quad f(\zeta_1)(x_1 - x_0) + f(\zeta_2)(x_2 - x_1) + \ldots + f(\zeta_n)(x_n - x_{n-1}).$$

The terms which follow the first in (3) are fixed. Since $f(x)$ is not bounded in (x_0, x_1), we can make $f(\zeta_1)$ large at pleasure by choosing ζ_1 appropriately. Then $f(\zeta_1)(x_1 - x_0)$ can be made arbitrarily large. Hence the sum (3) can be made arbitrarily large in absolute value by a proper choice of ζ_1. If the sum exceeds $|I| + \varepsilon$ in absolute value, it will differ from I by more than ε.

Q.E.D.

CONDITION FOR INTEGRABILITY

3. *Theorem: For $f(x)$ to be integrable in (a, b), it is necessary and sufficient that for every $\varepsilon > 0$ a $\delta > 0$ exist such that any two sums of norms less than δ have a difference which is less than ε in absolute value.*

This theorem is analogous to the fundamental theorem on convergent sequences.

Proof: A. *Necessity* - Let I exist. Take δ such that any sum of norm less than δ differs from I by less than $\frac{\varepsilon}{2}$. Then any two such sums differ by less than ε.

B. *Sufficiency* - Let

$$\varepsilon_1, \ \varepsilon_2 < \varepsilon_1, \ \varepsilon_3 < \varepsilon_2, \ldots, \ \varepsilon_{p+1} < \varepsilon_p, \ldots$$

be an infinite sequence of positive numbers which decrease toward zero.

For every p, chose a $\delta_p > 0$ such that any two sums of norms less than δ_p differ by less than ε_p. Furthermore, choose the δ's so that $\delta_1 > \delta_2 > \ldots > \delta_p > \ldots$, and so that the sequence of δ's tends toward zero.

For every p, form a sum Σ_p of norm less than δ_p. Consider the sequence

$$(4) \quad \Sigma_1, \Sigma_2, \ldots, \Sigma_p, \ldots$$

We say that this sequence approaches a limit.

Given any $\varepsilon > 0$, take p so that $\varepsilon_p < \varepsilon$. Then two elements of the sequence (4) beyond the pth differ by less than ε, for their norms are less than δ_p. Hence the sequence (4) approaches a limit.

Let I be the limit of the sequence (4).

Given any $\varepsilon > 0$, let $\delta > 0$ be chosen so that any two sums of norms less than δ differ by less than $\frac{\varepsilon}{2}$.

We say that any sum Σ of norm less than δ differs from I by less than ε.

For, let p be taken so that

$$|I - \Sigma_p| < \frac{\varepsilon}{2} \text{ and } \delta_p < \delta,$$

where Σ_p is as in (4) and δ_p is as above. Then

$$|I - \Sigma| = |(I - \Sigma_p) + (\Sigma_p - \Sigma)| < \frac{\varepsilon}{2} + \frac{\varepsilon}{2} = \varepsilon.$$

This proves the sufficiency of the condition.

INTEGRABILITY IN SUBINTERVALS

4. Theorem: *Let $f(x)$ be integrable in (a, b). Let (c, d) be any subinterval of (a, b). Then $f(x)$ is integrable in (c,d).*

Proof: It will suffice to consider any subinterval (a, c). Given an $\epsilon > 0$ let $\delta > 0$ be so chosen that two sums of norms less than δ, constructed for the interval (a, b), differ by less than ϵ.

Consider any two sums Σ_1 and Σ_2, formed for (a, c), of norms less than δ. We say that $|\Sigma_1 - \Sigma_2| < \epsilon$.

For, let (c, b) be divided in any way, into intervals of lengths less than δ. Let points ζ be chosen in these intervals.

Using this division of (c, b), and the subintervals and ζ's which appear in Σ_1, we form a sum S_1 for (a, b). Similarly, using the division of (c, b) and the subintervals and ζ's which appear in Σ_2, we form a sum S_2 for (a, b).

Now, as $|S_1 - S_2| < \epsilon$ and as $S_1 - S_2 = \Sigma_1 - \Sigma_2$, we have
$$|\Sigma_1 - \Sigma_2| < \epsilon.$$
Q.E.D.

5. Theorem: *Let $f(x)$ be integrable in (a, b). Let c be any point interior to (a, b). Then*
$$\int_a^b f(x)\ dx = \int_a^c f(x)\ dx + \int_c^b f(x)\ dx.$$

Proof: Choose a sequence of sums approaching $\int_a^b$, with norms tending toward zero and with c one of the points of subdivision for each sum. Let the sequence of sums be
$$\Sigma_1, \Sigma_2, \ldots, \Sigma_p, \ldots.$$

Let Σ_p' be that part of Σ_p which comes from intervals in (a, c) and Σ_p'' that part of Σ_p which comes from intervals in (c, b). Then
$$\Sigma_p = \Sigma_p' + \Sigma_p''.$$
The limit of Σ_p' is $\int_a^c$, that of Σ_p'' is $\int_c^b$.

Hence, $\int_a^b = \int_a^c + \int_c^b$.

BOUNDS FOR AN INTEGRAL

6. Theorem: *Let $f(x)$ be integrable in (a, b). Let M and m be respectively the least upper and greatest lower bounds of $f(x)$ in (a, b). Then*
$$m(b - a) \leq \int_a^b f(x)\ dx \leq M(b - a).$$

Proof: We limit ourselves to the second inequality. For any sum, we have
$$f(\zeta_1)(x_1 - x_0) + \ldots + f(\zeta_n)(x_n - x_{n-1}) \leq M(x_1 - x_0) + \ldots + M(x_n - x_{n-1})$$
$$= M(x_n - x_0) = M(b - a).$$

Then the integral cannot exceed $M(b - a)$. If it did, any sum of sufficiently small norm would exceed $M(b - a)$.

Theorem: *Let $f(x)$ be integrable in (a, b). If $|f(x)| \leq G$ in (a, b), then $\left|\int_a^b f(x)\ dx\right| \leq G(b - a)$.*

Proof: Clear.

CONTINUITY OF INTEGRAL WITH RESPECT TO UPPER LIMIT

7. Let $f(x)$ be integrable in (a, b). Let $\underline{x}$ be any point of (a, b). Then, if $x > a$,
$$\int_a^x f(x)\,dx$$
exists. For $a = a$, we *define* $\int_a^x f(x)\,dx$ as *zero*. The integral $\int_a^x f(x)\,dx$ is a function of the upper limit x. We designate this function by $F(x)$.

We shall prove that the function $F(x)$ is continuous throughout (a, b).

Case A. Let $\underline{h}$ be small and positive. Then
$$F(x + h) = F(x) + \int_x^{x+h} f(x)\,dx.$$
Hence
$$|F(x + h) - F(x)| = \left|\int_x^{x+h} f(x)\,dx\right|.$$
Let $|f(x)| \leq G$ in (a, b). Then
$$|F(x + h) - F(x)| \leq Gh$$
and Gh is small when $\underline{h}$ is small.

Case B. Let $\underline{h}$ be small and negative. Then
$$F(x + h) = F(x) - \int_{x+h}^x f(x)\,dx$$
and
$$|F(x + h) - F(x)| = \left|\int_{x+h}^x f(x)\,dx\right| \leq G(-h).$$

Q.E.D.

INTEGRABILITY OF SUM OF TWO FUNCTIONS

8. *Theorem:* Let $f(x)$ *and* $g(x)$ *be integrable in* (a, b). *Then* $f(x) + g(x)$ *is integrable in* (a, b) *and*
$$\int_a^b [f(x) + g(x)]\,dx = \int_a^b f(x)\,dx + \int_a^b g(x)\,dx.$$

Proof: Given an $\varepsilon > 0$, let $\delta > 0$ be taken so that any sum formed for $f(x)$, of norm less than δ, differs from $\int_a^b f(x)\,dx$ by less than $\frac{\varepsilon}{2}$; also so that any sum for $g(x)$ of norm less than δ differs from $\int_a^b g(x)\,dx$ by less than $\frac{\varepsilon}{2}$. Consider any sum of norm less than δ for $f(x) + g(x)$. Let it be
$$\Sigma = [f(\zeta_1) + g(\zeta_1)](x_1 - x_0) + \ldots + [f(\zeta_n) + g(\zeta_n)](x_n - x_{n-1}).$$
Now $\Sigma = \Sigma_1 + \Sigma_2$, where
$$\Sigma_1 = f(\zeta_1)(x_1 - x_0) + \ldots + f(\zeta_n)(x_n - x_{n-1})$$
$$\Sigma_2 = g(\zeta_1)(x_1 - x_0) + \ldots + g(\zeta_n)(x_n - x_{n-1}).$$
As
$$\left|\int_a^b f(x)\,dx - \Sigma_1\right| < \frac{\varepsilon}{2} \text{ and } \left|\int_a^b g(x)\,dx - \Sigma_2\right| < \frac{\varepsilon}{2},$$
we have
$$\left|\int_a^b f(x)\,dx + \int_a^b g(x)\,dx - \Sigma\right| < \varepsilon.$$

Q.E.D.

INTEGRABILITY OF CONTINUOUS FUNCTIONS

9. *Theorem:* Let $f(x)$ *be continuous throughout* (a, b). *Then* $f(x)$ *is integrable in* (a, b).

Proof: We know that $f(x)$ is uniformly continuous in (a, b). Given an $\varepsilon > 0$, let $\delta > 0$ be taken so that, if $|x' - x| < \delta$, with x and x' in (a, b), we have

$$|f(x') - f(x)| < \frac{\varepsilon}{2(b - a)}.$$

We say that any two sums Σ_1 and Σ_2 of norms less than δ differ by less than ε. This will prove our theorem.

Take all of the points of division which figure in Σ_1 and all which figure in Σ_2. The totality of these points gives a new division of (a, b). Let Σ_3 be any sum based on this third division. We shall study the difference between Σ_3 and Σ_1.

Let (p, q) be any interval which is used in forming Σ_1. Then (p, q) decomposes, when the points of division used for Σ_2 are added, into intervals

$$(x_0, x_1), (x_1, x_2), \ldots, (x_{n-1}, x_n).$$

Let ζ be the point chosen for (p, q) in forming Σ_1. In forming Σ_3, let ζ_i be chosen in (x_{i-1}, x_i), $i = 1, \ldots, n$.

The terms of Σ_3 which come out of (p, q) have a sum

$$A = f(\zeta_1)(x_1 - x_0) + \ldots + f(\zeta_n)(x_n - x_{n-1}).$$

As to Σ_1, we get from (p, q) a single item,

$$B = f(\zeta)(q - p) = f(\zeta)(x_1 - x_0) + \ldots + f(\zeta)(x_n - x_{n-1}).$$

Then

$$A - B = (x_1 - x_0)[f(\zeta_1) - f(\zeta)] + \ldots + (x_n - x_{n-1})[f(\zeta_n) - f(\zeta)].$$

As $|\zeta_i - \zeta| < \delta$ for every i, we have

$$|f(\zeta_i) - f(\zeta)| < \frac{\varepsilon}{2(b - a)}$$

for every i.

Hence

$$|A - B| < \frac{\varepsilon}{2(b - a)}[(x_1 - x_0) + \ldots + (x_n - x_{n-1})] = \frac{\varepsilon}{2(b - a)}(q - p).$$

It follows easily from this that

$$|\Sigma_3 - \Sigma_1| < \frac{\varepsilon}{2(b - a)}(b - a) = \frac{\varepsilon}{2}.$$

Likewise, $|\Sigma_3 - \Sigma_2| < \frac{\varepsilon}{2}$, so that $|\Sigma_2 - \Sigma_1| < \varepsilon$.

Q.E.D.

DIFFERENTIABILITY OF THE INTEGRAL OF A CONTINUOUS FUNCTION

10. Theorem: *Let $f(x)$ be integrable in (a, b) and continuous at some point x_0 of (a, b). Let $F(x) = \int_a^x f(x)\,dx$. Then $F(x)$ has a derivative at x_0 and its derivative at x_0 equals $f(x_0)$.*

Proof: Case A. Let $h > 0$. Then

$$F(x_0 + h) - F(x_0) = \int_{x_0}^{x_0+h} f(x)\,dx.$$

Let M be the least upper bound of $f(x)$ in (x_0, x_0+h) and m the greatest lower bound. Because of the continuity of $f(x)$ at x_0, given any $\varepsilon > 0$, a $\delta > 0$ can be found such that

$$M - f(x_0) < \varepsilon, \qquad f(x_0) - m < \varepsilon$$

for $h < \delta$.

By §6,
$$mh \leq F(x_0 + h) - F(x_0) \leq Mh$$
so that
$$m \leq \frac{F(x_0 + h) - F(x_0)}{h} \leq M.$$

We see now easily from (1) that
$$\left|\frac{F(x_0 + h) - F(x_0)}{h} - f(x_0)\right| < \varepsilon.$$

Case B. $h < 0$.

Proof: Similar to above.

Q.E.D.

EVALUATION OF DEFINITE INTEGRALS

11. *Theorem: Let $f(x)$ be continuous on the closed interval (a, b). Let $G(x)$ be any function differentiable on (a, b) such that*
$$\frac{dG(x)}{dx} = f(x)$$
for each x on (a, b). Then

(1) $$\int_a^b f(x)\, dx = G(b) - G(a).$$

Proof: We use $F(x)$ as in §10. Then $F(x)$ is differentiable on (a, b) and
$$\frac{d}{dx}[F(x) - G(x)] = 0$$
throughout (a, b). Thus a constant c exists such that
$$F(x) = G(x) + c.$$
As $F(a) = 0$, we have $c = -G(a)$. Thus
$$F(b) = G(b) - G(a).$$
As $F(b)$ is the definite integral in (1), the theorem is proved.

Q.E.D.

COMPOSITION OF INTERVALS

12. *Theorem: Let $a < b < c$. Let $f(x)$ be integrable in (a, b) and in (b, c). Then $f(x)$ is integrable in (a, c).*

Proof: Let $G > 0$ be such that $|f(x)| < G$ in (a, b) and in (b, c); that is, $|f(x)| < G$ throughout (a, c).

Let $\varepsilon > 0$ be assigned. Take δ so that

(1) $0 < \delta < \frac{\varepsilon}{6G}$;

(2) Any sum for (a, b) of norm less than δ differs from $\int_a^b f(x)\, dx$ by less than $\frac{\varepsilon}{3}$;

(3) Any sum for (b, c) of norm less than δ differs from $\int_b^c f(x)\, dx$ by less than $\frac{\varepsilon}{3}$.

We say that any sum for (a, c) of norm less than δ differs from $\int_a^b + \int_b^c$ by less than ε.

Let (a, c) be divided into intervals of length less than δ. Let Σ be a sum formed for this division.

If b is a point of division, we surely have our result.

Suppose that b is not a point of division Let b be contained in the interval (p, q) of the division. Let ζ be the point chosen in (p, q).

When we add b to the points of division, we secure a division of (a, b) and of (b, c). Choose a point ζ_1 in (p, b) and a ζ_2 in (b, q).

Form a sum Σ_1 for (a, b), using ζ_1 for (p, b) and, for the other ζ's, those which appear in Σ.

Form a sum Σ_2 for (b, c), using ζ_2 for (b, q) and, for the other ζ's, those which appear in Σ.

Then
$$\Sigma - (\Sigma_1 + \Sigma_2) = f(\zeta)(q - p) - f(\zeta_1)(b - p) - f(\zeta_2)(q - b)$$
$$= [f(\zeta) - f(\zeta_1)](b - p) + [f(\zeta) - f(\zeta_2)](q - b),$$

so that
$$|\Sigma - (\Sigma_1 + \Sigma_2)| < 2G(b - p) + 2G(q - b) = 2G(q - p) < 2G\delta < 2G \frac{\varepsilon}{6G} = \frac{\varepsilon}{3}.$$

Hence
$$|\Sigma - (\int_a^b + \int_b^c)| < \varepsilon.$$

Q.E.D.

DISCONTINUOUS FUNCTIONS

13. *Theorem: Let f(x) be bounded in (a, b), and continuous, except perhaps at a finite number of points. Then f(x) is integrable in (a, b).*

Proof: By virtue of §12 it will suffice to consider the case in which f(x) is continuous, expect perhaps at one extremity.

We assume then that f(x) is continuous in (a, b) except perhaps at b.

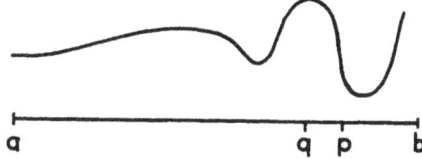

Let $|f(x)| < G$ in (a, b).

Let $\varepsilon > 0$ be assigned.

Choose a point p, interior to (a, b) so that $b - p < \frac{\varepsilon}{12G}$.

Take δ so that any sum for (a, p) of norm less than δ differs from $\int_a^p f(x)\,dx$ by less than $\frac{\varepsilon}{6}$; and also so that $\delta < \frac{\varepsilon}{12G}$.

Take any division of (a, b) into intervals of length less than δ. Let Σ be a sum for this division. Let q be the rightmost point of division which is not to the right of p. Then
$$b - q < \frac{2\varepsilon}{12G} = \frac{\varepsilon}{6G}.$$

We shall prove that that part of Σ which comes from (a, q) differs from $\int_a^q$ by not more than $\frac{\varepsilon}{6}$. Suppose that the absolute value of the difference exceeds $\frac{\varepsilon}{6}$; let the absolute value be $\frac{\varepsilon}{6} + \eta$ with $\eta > 0$.

Form a sum for (q, p) with intervals so small that this sum differs from $\int_q^p$ by less than η. (If q = p, the sum, by definition, is to be zero.) Then this sum, plus the part of Σ which comes from (a, q), gives a sum for (a, p), of norm less than δ, differing from $\int_a^p$ by more than $\frac{\varepsilon}{6}$. This contradicts the nature of δ.

That part of Σ which comes from (q, b) has an absolute value no greater than
$$G \cdot \frac{\varepsilon}{6G} = \frac{\varepsilon}{6}.$$

Hence Σ differs from $\int_a^q$ by no more than $\frac{\varepsilon}{3}$. Finally, $\int_a^q$ differs from $\int_a^P$ by $\int_q^P$, the absolute value of which is no more than $G \frac{\varepsilon}{12G} = \frac{\varepsilon}{12}$.

Hence Σ differs from $\int_a^P$ by less than $\frac{\varepsilon}{2}$.

Thus, if Σ_1 and Σ_2 are two sums for (a, b) of norms less than δ, they will each differ from $\int_a^P$ by less than $\frac{\varepsilon}{2}$, and will differ from each other by less than ε.

Q.E.D.

VII

Infinite Series of Numbers

INFINITE SERIES

1. Let $u_1, u_2, \ldots, u_n, \ldots$ be any infinite sequence of real numbers. The symbol

(1) $$u_1 + u_2 + \ldots + u_n + \ldots$$

will be called an infinite *series* of real numbers.

We sometimes use the symbol

$$\sum_{i=1}^{\infty} u_i$$

to represent the series (1).

CONVERGENT INFINITE SERIES

2. Let

(2) $$u_1 + u_2 + \ldots + u_n + \ldots$$

be any infinite series.

Let $s_1 = u_1$

$s_2 = u_1 + u_2$

$s_3 = u_1 + u_2 + u_3$

$\ldots\ldots\ldots\ldots\ldots\ldots\ldots$

$s_n = u_1 + u_2 + \ldots + u_n.$

Consider the sequence

(3) $$s_1, s_2, \ldots, s_n, \ldots.$$

If (3) converges, that is, if it has a limit, s, we shall say that (2) converges and we shall call s the *sum* of (2).

If (3) diverges, that is, if it has no limit, we shall say that (2) *diverges*.

Examples: The series

$$\frac{1}{2} + \frac{1}{4} + \ldots + \frac{1}{2^n} + \ldots$$

converges. Its sum is 1. For the sequence

$$\frac{1}{2}, \frac{3}{4}, \ldots, 1 - \frac{1}{2^n}, \ldots$$

converges and its limit is 1.

The two series

$$1 + 1 + \ldots + 1 + \ldots$$

and
$$1 - 1 + 1 - 1 + \ldots + (-1)^{n+1} + \ldots$$
are divergent.

A NECESSARY AND SUFFICIENT CONDITION FOR CONVERGENCE

3. *Theorem: For $u_1 + \ldots + u_n + \ldots$ to converge, it is necessary and sufficient that for every $\varepsilon > 0$, a positive integer N exist such that, for every $n > N$ and for every positive integer p,*
$$|u_{n+1} + \ldots + u_{n+p}| < \varepsilon.$$

Proof: The associated sequence
$$s_1, s_2, \ldots, s_n, \ldots$$
is such that
$$s_{n+p} - s_n = u_{n+1} + \ldots + u_{n+p}$$
The theorem then follows immediately from the fundamental theorem on convergent sequences.

ADDITION OF CONVERGENT SERIES

4. *Theorem: If*

(4) $\quad u_1 + u_2 + \ldots + u_n + \ldots$

and

(5) $\quad v_1 + v_2 + \ldots + v_n + \ldots$

converge, then
$$(u_1 + v_1) + (u_2 + v_2) + \ldots + (u_n + v_n) + \ldots$$
converges and its sum is the sum of the sums of (4) and (5).

Proof: Clear.

THE nTH TERM

5. It follows from §3 that if $u_1 + \ldots + u_n + \ldots$ converges then u_n approaches zero as n increases indefinitely.

The converse is not true. For instance, the harmonic series
$$1 + \frac{1}{2} + \frac{1}{3} + \ldots + \frac{1}{n} + \ldots$$
diverges, even though its nth term approaches zero.

THE REMAINDER

6. Let

(6) $\quad u_1 + u_2 + \ldots + u_n + \ldots$

converge to a sum s. Clearly, for every n, the series
$$u_{n+1} + u_{n+2} + \ldots$$
converges, and its sum is $s - (u_1 + \ldots + u_n) = s - s_n$.

We shall call the sum of the series (7), that is, $s - s_n$, the *remainder after n terms* in the series (6).

Clearly, the remainder after n terms, that is, $s - s_n$, approaches 0 as n increases indefinitely.

We have
$$s = s_n + R_n$$
where R_n is the remainder after n terms.

A CLASS OF CONVERGENT SERIES

7. Let $a_1, a_2, \ldots, a_n, \ldots$ be an infinite sequence of positive numbers which is such that

(a) $a_1 > a_2 > a_3 > \ldots > a_n > \ldots$

(b) a_n approaches zero as n increases indefinitely.

We say that the *series*

(8) $\qquad a_1 - a_2 + a_3 - a_4 + \ldots + (-1)^{n+1} a_n + \ldots$

converges

Example: $1 - \frac{1}{2} + \frac{1}{3} - \frac{1}{4} + \ldots$ converges.

Proof: We study the sequence

(9) $\qquad s_1, s_2, \ldots, s_n, \ldots$

Take the subsequence of (9)

(10) $\qquad s_3, s_5, \ldots, s_{2n+1}, \ldots$

We show first that (10) converges, then that (9) converges to the limit to which (10) does.

In accordance with the fundamental theorem on sequences, we have to consider the difference
$$s_{2(n+p)+1} - s_{2n+1}.$$

Now
$$s_{2n+1} = a_1 - a_2 + \ldots + a_{2n+1}$$
$$s_{2(n+p)+1} + a_1 - a_2 + \ldots + a_{2n+1} - a_{2n+2} + \ldots + a_{2(n+p)+1}.$$

Hence
$$s_{2(n+p)+1} - s_{2n+1} = (a_{2(n+p)+1} - a_{2(n+p)}) + \ldots + (a_{2n+3} - a_{2n+2}),$$
and, as each parenthesis in the last equation is negative, we have

(11) $\qquad s_{2(n+p)+1} - s_{2n+1} < 0$

On the other hand, we have
$$s_{2(n+p)+1} - s_{2n+1} = a_{2(n+p)+1} - (a_{2(n+p)} - a_{2(n+p)-1}) - \ldots - (a_{2n+4} - a_{2n+3}) - a_{2n+2},$$
so that, since each parenthesis is again negative, we have

(12) $\qquad s_{2(n+p)+1} - s_{2n+1} > a_{2(n+p)+1} - a_{2n+2} > -a_{2n+2}.$

From (11) and (12), we have

$$|s_{2(n+p)+1} - s_{2n+1}| < a_{2n+2},$$

so that, since a_{an+2} tends toward zero as n increases, (10) converges.

Let s be the limit of (10). We say that *(9) converges to s*.

We have $s_{2n+1} - s_{2n} = a_{2n+1}$.

Let $\varepsilon > 0$ be assigned. If N is so large that, for $2n + 1 > N$ we have $a_{2n+1} < \frac{\varepsilon}{2}$ and $|s - s_{2n+1}| < \frac{\varepsilon}{2}$ then, for $2n > N$, we have $|s - s_{2n}| < \varepsilon$. That is, every sum s_i with $i > N$ differs from s by less than ε.

Q.E.D.

SERIES OF ABSOLUTE VALUES

8. *Theorem: Let*

(12) $u_1 + u_2 + \ldots + u_n + \ldots$

be an infinite series. If

(13) $|u_1| + |u_2| + \ldots + |u_n| + \ldots$

converges, then (12) also converges.

Proof: We have

$$|u_{n+1} + \ldots + u_{n+p}| \leq |u_{n+1}| + \ldots + |u_{n+p}|.$$

Q.E.D.

Remark: The converse is not true. For instance, $1 - \frac{1}{2} + \frac{1}{3} - \frac{1}{4} + \ldots$ converges, but not $1 + \frac{1}{2} + \frac{1}{3} + \frac{1}{4} + \ldots$.

ABSOLUTE CONVERGENCE

9. *Definition:* Let

(14) $u_1 + \ldots + u_n + \ldots$

be a convergent infinite series. If

(15) $|u_1| + \ldots + |u_n| + \ldots$

converges, we shall call (14) *absolutely convergent*. If (15) diverges, we shall call (14) *semi-convergent*.

(If (14) is absolutely convergent, the sum of (14) cannot exceed the sum of (15). For the difference between (15) and (14) is

$$[|u_1| - u_1] + \ldots + [|u_n| - u_n] + \ldots$$

whose terms are non-negative.)

REARRANGEMENT OF TERMS OF AN ABSOLUTELY CONVERGENT SERIES

10. Let

(16) $u_1 + \ldots + u_n + \ldots$

be absolutely convergent. Let

(17) $v_1 + \ldots + v_n + \ldots$

be the series (16) with its terms rearranged in any order. That is, the series (17) is a series

$$u_{i_1} + u_{i_2} + \ldots + u_{i_n} + \ldots$$

where $i_1, i_2, \ldots$ constitute a rearrangement of the positive integers $1, 2, \ldots$.

We say that *(17) converges absolutely, and that its sum is the sum of (16).*

Proof: Consider the convergent series

(18) $$|u_1| + \ldots + |u_n| + \ldots$$

Let R_n be the remainder after n terms in (18). Let $\varepsilon > 0$ be assigned and let $|R_n| < \varepsilon$ for $n \geq N$. Let M be so large that the sum

$$v_1 + v_2 + \ldots + v_M$$

contains all of the terms $u_1, u_2, \ldots, u_N$. Let σ_n represent the sum of the first n terms of (17).

For $n > M$, the difference between σ_n and the sum of the first N terms of (16) is a sum of terms of (16) which follows the Nth. The sum of any finite number of terms of (16) which follow the Nth does not exceed the remainder after N terms in (18). Hence the difference between σ_n and the sum of the first N terms in (16) is less than ε in absolute value. But the sum of the first N terms of (16) differs from the sum of (16) by no more than the remainder after N terms in (18). Hence, σ_n with $n > M$ differs from the sum of (16) by less than 2ε. This proves that (17) converges to the sum of (16).

The series $|v_1| + \ldots + |v_n| + \ldots$ is a rearrangement of (18).

Hence (17) is absolutely convergent. Q.E.D.

11. If

$$u_1 + u_2 + \ldots + u_n + \ldots$$

is only semi-convergent, we can, by properly rearranging its terms, get a series which converges to any desired limit.

VIII

Sequences of Functions

CONVERGENT SEQUENCES OF FUNCTIONS

1. Let

$$(1) \qquad s_1(x), s_2(x), \ldots, s_n(x), \ldots$$

be a sequence of functions which are defined on a domain E.

Suppose that (1) converges for every value of x in E.

For every x of E, let the limit of (1) be represented by $s(x)$. We have, in $s(x)$, a function defined on E. We shall call $s(x)$ the *limit* of the sequence (1).

UNIFORM CONVERGENCE

2. We shall say that the sequence (1) converges *uniformly* on E if, for every $\varepsilon > 0$, we can determine an N such that, for $n > N$,

$$|s(x) - s_n(x)| < \varepsilon$$

for every x of E.

Very plainly, if E consists of a finite number of points, convergence on E implies uniform convergence on E.

EXAMPLE OF A UNIFORMLY CONVERGENT SEQUENCE DEFINED ON AN INTERVAL

3. Let E be the closed interval $(0, 1)$. Let the sequence be

$$x^2 - x, \; x^2 - \frac{x}{2}, \; x^2 - \frac{x}{3}, \ldots, x^2 - \frac{x}{n}, \ldots$$

The limit, $s(x)$, is x^2. Now, for x in $(0, 1)$,

$$|s(x) - s_n(x)| = \frac{x}{n} \leq \frac{1}{n}.$$

If $\varepsilon > 0$ is assigned, take $N > \frac{1}{\varepsilon}$. For $n > N$,

$$|s(x) - s_n(x)| \leq \frac{1}{n} < \frac{1}{N} < \varepsilon,$$

for every x in $(0, 1)$.

EXAMPLE OF NON-UNIFORM CONVERGENCE

4. Let E be the closed interval $(0, 1)$. Define $s_n(x)$ as follows:

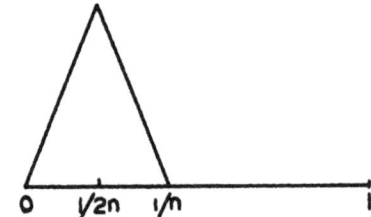

For $0 \leq x \leq \frac{1}{2n}$, $\quad s_n(x) = 2nx;$

For $\frac{1}{2n} < x \leq \frac{1}{n}$, $\quad s_n(x) = 2n\left(\frac{1}{n} - x\right);$

For $\frac{1}{n} < x \leq 1$, $\quad s_n(x) = 0.$

Then $s_n(x)$ approaches 0 from every x. This is immediately obvious for $x = 0$. For $x > 0$, we have $s_n(x) = 0$ for n sufficiently large; to be precise, for $n > \frac{1}{x}$. Hence the sequence $s_1(x),\ldots, s_n(x),\ldots$ converges to zero.

But the convergence is not uniform, for

$$|s(\tfrac{1}{2n}) - s_n(\tfrac{1}{2n})| = |0 - 1| = 1$$

Hence, if we take $\varepsilon = 1$, we cannot find an N such that $|s(x) - s_n(x)| < \varepsilon$ for $n > N$ and for every x.

NECESSARY AND SUFFICIENT CONDITION FOR THE UNIFORM CONVERGENCE OF A SEQUENCE

5. *Theorem:* For

(2) $\qquad s_1(x), s_2(x),\ldots, s_n(x),\ldots$

to converge uniformly on a set E, it is necessary and sufficient that for every $\varepsilon > 0$ a positive integer N exist such that, for every $n > N$ and for every positive integer p,

(3) $\qquad |s_{n+p}(x) - s_n(x)| < \varepsilon$

for every x on E.

Proof:

A. *Necessity* - Let $s(x)$ be the limit of (1). If N is taken so that $|s(x) - s_n(x)| < \frac{\varepsilon}{2}$ throughout E for $n > N$, then we have (3).

Q.E.D.

B. *Sufficiency* - Let the condition be satisfied. Of course, (2) will converge throughout E. An $\varepsilon > 0$ being given, let N be taken so that (3) holds throughout E. We say that, for $n > N$,

(4) $\qquad |s(x) - s_n(x)| \leqq \varepsilon$

on E. This will prove uniform convergence.

Let x_0 be a point for which (4) does not hold, that is for which

$$|s(x_0) - s_n(x_0)| > \varepsilon$$

for some $n > N$. Keep n fixed. It is clear that, for p very large, we have

$$|s_{n+p}(x_0) - s_n(x_0)| > \varepsilon$$

because $s_{n+p}(x_0)$ is an arbitrarily good approximation to $s(x_0)$ when p is large.

Q.E.D.

CONTINUITY OF LIMIT OF A SEQUENCE

6. *Theorem:* Let

$$s_1(x), s_2(x),\ldots, s_n(x),\ldots$$

converge uniformly on a closed interval (a, b). Let the functions of the sequence all be continuous at some point x_0 of (a, b). Then $s(x)$, the limit of the sequence, is continuous at x_0.

Remarks: If the functions $s_n(x)$ are continuous throughout (a, b), then $s(x)$ is continuous throughout (a, b). The assumption that (a, b) is closed is not essential. It is made in order to take care of the case in which x_0 is an end-point of the interval.

Proof: An $\varepsilon > 0$ being assigned, take n so that $|s(x) - s_n(x)| < \frac{\varepsilon}{3}$ throughout (a, b). Keep n fixed. As $s_n(x)$ is continuous at x_0, there is a $\delta > 0$ such that, for $|x - x_0| < \delta$, we have

$$|s_n(x) - s_n(x_0)| < \varepsilon.$$

Then, for $|x - x_0| < \delta$,

$$|s(x) - s(x_0)| \leq |s(x) - s_n(x)| + |s_n(x) - s_n(x_0)| + |s_n(x_0) - s(x_0)| < \frac{\varepsilon}{3} + \frac{\varepsilon}{3} + \frac{\varepsilon}{3} = \varepsilon.$$

Q.E.D.

EXAMPLE OF A SEQUENCE OF CONTINUOUS FUNCTIONS WHOSE LIMIT IS NOT CONTINUOUS

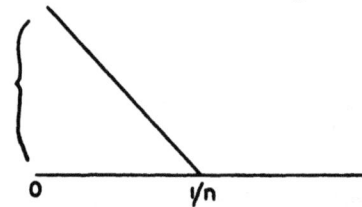

7. We use the closed interval $(0, 1)$.

We take

$$s_n(x) = 1 - nx \text{ for } 0 \leq x < \frac{1}{n},$$
$$s_n(x) = 0 \text{ for } \frac{1}{n} \leq x \leq 1.$$

If $x > 0$, $s_n(x) = 0$ for $n > \frac{1}{x}$, so that $s(x) = 0$. On the other hand, $s_n(0) = 1$ for every n, so that $s(0) = 1$. Thus, $s(x)$ is discontinuous at 0.

To see that the sequence is not uniformly convergent, we observe that $s_n(\frac{1}{2n}) = \frac{1}{2}$ for every n. Hence, for every n,

$$|s(\frac{1}{2n}) - s_n(\frac{1}{2n})| = |0 - \frac{1}{2}| = \frac{1}{2},$$

and, if we are given $\varepsilon = \frac{1}{2}$, we cannot find an N which will give $|s(x) - s_n(x)| < \varepsilon$ throughout $(0, 1)$ for $n > N$.

INTEGRABILITY OF LIMITS OF SEQUENCES

8. *Theorem: Let (a, b) be a closed interval. Let*

$$s_1(x), s_2(x), \ldots, s_n(x), \ldots$$

be a sequence of functions each one of which is integrable on (a, b). Let the sequence converge uniformly on (a, b) to a limit $s(x)$. Then

(1) *$s(x)$ is integrable;*
(2) *$\int_a^b s_1(x)\, dx, \ldots, \int_a^b s_n(x)\, dx, \ldots$ is a convergent sequence;*
(3) *$\int_a^b s(x)\, dx = \lim\limits_{n \to \infty} \int_a^b s_n(x)\, dx.$*

Proof: To prove (1), we have to show that two sums for $s(x)$ which are of small norms have a small difference.

Let

$$R_n(x) = s(x) - s_n(x).$$

Let $\varepsilon > 0$ be assigned. Take n so that $|R_n(x)| < \varepsilon$ for every x on (a, b). Keep n fixed. Take δ so that any two sums for $s_n(x)$ of norms less than δ differ by less than ε.

We have $s(x) = s_n(x) + R_n(x)$.

Hence

$$\Sigma = s(\zeta_1)(x_1 - x_0) + \ldots + s(\zeta_p)(x_p - x_{p-1})$$

and

$$\Sigma' = s_n(\zeta_1)(x_1 - x_0) + \ldots + s_n(\zeta_p)(x_p - x_{p-1}),$$

we have

$$\Sigma - \Sigma' = R_n(\zeta_1)(x_1 - x_0) + \ldots + R_n(\zeta_p)(x_p - x_{p-1})$$

and
$$|\Sigma - \Sigma'| < \varepsilon(x_1 - x_0) + \ldots + \varepsilon(x_p - x_{p-1}) = \varepsilon(b - a).$$

As any two sums for $s_n(x)$ of norms less than δ differ by less than ε, any two such sums for $s(x)$ differ by less than $\varepsilon + 2\varepsilon(b - a)$, which is small when ε is small. Hence $s(x)$ is integrable.

It remains to prove (2) and (3).

We consider the difference
$$\int_a^b s(x)\,dx - \int_a^b s_n(x)\,dx.$$

We know that $s(x) - s_n(x)$ is integrable and that
$$\int_a^b s(x)\,dx - \int_a^b s_n(x)\,dx = \int_a^b [s(x) - s_n(x)]\,dx.$$

An $\varepsilon > 0$ being assigned, take N so that $|s(x) - s_n(x)| < \varepsilon$ for $n > N$ throughout (a, b). Then
$$\left|\int_a^b s(x)\,dx - \int_a^b s_n(x)\,dx\right| = \left|\int_a^b [s(x) - s_n(x)]\,dx\right| \leq \varepsilon\,|b - a|.$$

As $\varepsilon\,|b - a|$ is small when ε is small, we see the truth of (2) and (3).

9. When the sequence does not converge uniformly, $s(x)$ may not be integrable, or the sequence of integrals may diverge, or $s(x)$ may be integrable and the sequence of integrals may not converge to $s(x)$.

We give an example to show that the sequence of integrals may converge to a limit different from the integral of $s(x)$.

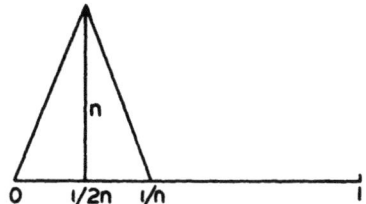

Let
$$s_n(x) = 2n^2 x \text{ for } 0 \leq x \leq \tfrac{1}{2n},$$
$$s_n(x) = 2n^2(\tfrac{1}{n} - x) \text{ for } \tfrac{1}{2n} \leq x \leq \tfrac{1}{n},$$
$$s_n(x) = 0 \text{ for } \tfrac{1}{n} \leq x \leq 1.$$

Then $s(x) = 0$ for every $\underline{x}$. As $\int_0^1 s(x)\,dx = 0$ and $\int_0^1 s_n(x)\,dx = \tfrac{1}{2}$ for every $\underline{n}$, we see that the limit of the integrals of the functions $s_n(x)$ is not the integral of the limit.

DIFFERENTIABILITY OF SEQUENCES OF FUNCTIONS

10. *Theorem: Let*

(5) $\qquad s_1(x), \ldots, s_n(x), \ldots$

be a sequence of functions each of which has a derivative, $s_n'(x)$ throughout the closed interval (a, b). Let (5) converge on (a, b). Let

(6) $\qquad s_1'(x), \ldots, s_n'(x), \ldots$

converge uniformly on (a, b). Then

(a) *The sequence (5) converges uniformly on (a, b).*

(b) *The limit of (5) is differentiable throughout (a, b).*

(c) *The derivative of the limit of (5) is the limit of the sequence (6).*

Proof: (a) Given an $\varepsilon > 0$, take $\underline{N}$ so that
$$|s_{n+p}'(x) - s_n'(x)| < \varepsilon$$
throughout (a, b) for $n > N$ and also so that

$$|s_{n+p}(a) - s_n(a)| < \varepsilon$$

for $n > N$.

Let
$$\varphi(x) = s_{n+p}(x) - s_n(x).$$

Then
$$\varphi(x) - \varphi(a) = [s_{n+p}(x) - s_n(x)] - [s_{n+p}(a) - s_n(a)].$$

By the mean value theorem,
$$\varphi(x) - \varphi(a) = (x - a)\varphi'(x_1) = (x - a)[s'_{n+p}(x_1) - s'_n(x_1)]$$

where x_1 lies between a and x. Then

$$[s_{n+p}(x) - s_n(x)] = [s_{n+p}(a) - s_n(a)] + (x - a)[s'_{n+p}(x_1) - s'_n(x_1)]$$

Hence
$$|s_{n+p}(x) - s_n(x)| < \varepsilon + |(b - a)|\varepsilon$$

for every x on (a, b). This proves (a).

We attend now to (b) and (c).

We have, letting $R_n(x) = s(x) - s_n(x)$,

$$(7) \quad \frac{s(x + h) - s(x)}{h} = \frac{s_n(x + h) - s_n(x)}{h} + \frac{R_n(x + h) - R_n(x)}{h}$$

for $h \neq 0$.

Let $\varepsilon > 0$ be given. For $n > N$, let $|s'_{n+p}(x) - s'_n(x)| < \varepsilon$ throughout (a, b) for every p. Choose an $n > N$ and hold n fast.

We say that, for every x on (a, b) and for every admissible h,

$$(8) \quad \left|\frac{R_n(x + h) - R_n(x)}{h}\right| \leq \varepsilon.$$

Take any x and h and hold them fast. If we had

$$(9) \quad \left|\frac{R_n(x + h) - R_n(x)}{h}\right| > \varepsilon$$

we would have, for p very large,

$$(10) \quad \left|\frac{s_{n+p}(x + h) - s_n(x + h)}{h} - \frac{s_{n+p}(x) - s_n(x)}{h}\right| > \varepsilon.$$

[Note that the limit, as p increases, of the first member of (10) is the first member of (9).]

By the mean value theorem, we would have, for some x_1 between x and $x + h$,

$$|s'_{n+p}(x_1) - s'_n(x_1)| > \varepsilon$$

which is impossible. This proves (8).

Take δ so that $\left|\frac{s_n(x + h) - s_n(x)}{h} - s'_n(x)\right| < \varepsilon$ for $0 < |h| < \delta$. (We are dealing now with a fixed x.)

Then, by (7) and (8),

$$\left|\frac{s(x + h) - s(x)}{h} - s'_n(x)\right| < 2\varepsilon.$$

Let $\sigma(x)$ be the limit of (6). We have
$$|(s_n'(x) - \sigma(x)| \leq \varepsilon.$$
Otherwise $|s_n'(x) - s_{n+p}'(x)|$ would exceed ε for $\underline{p}$ large. Then $|\frac{s(x + h) - s(x)}{h} - \sigma(x)| < 3\varepsilon$ for $0 < |h| < \delta$. This proves (b) and (c).

IX

Infinite Series of Functions

INFINITE SERIES OF FUNCTIONS

1. Let $u_1(x), u_2(x), \ldots, u_n(x), \ldots$ be an infinite sequence of functions, all defined on some set E.

We call the symbol

$$(1) \quad u_1(x) + u_2(x) + \ldots + u_n(x) + \ldots$$

an *infinite series of functions*.

For every x of E, (1) becomes an infinite series of *numbers*.

With the series (1), we associate the infinite sequence of functions

$$(2) \quad s_1(x), s_2(x), \ldots, s_n(x), \ldots$$

where

$$s_n(x) = u_1(x) + \ldots + u_n(x).$$

We call (2) the *sequence associated with (1)*.

Every sequence

$$(3) \quad s_1(x), s_2(x), \ldots, s_n(x), \ldots$$

is a sequence associated with an infinite series. In short, (3) is associated with the series

$$s_1(x) + [s_2(x) - s_1(x)] + [s_3(x) - s_2(x)] + \ldots + [s_n(x) - s_{n-1}(x)] + \ldots$$

In this way, the theory of infinite series is made equivalent to the theory of infinite sequences.

UNIFORM CONVERGENCE OF INFINITE SERIES

2. If

$$(4) \quad u_1(x) + u_2(x) + \ldots + u_n(x) + \ldots$$

converges for every x on E, we say that (4) is *convergent on E*. Let (4) converge on E. By *the remainder after n terms of (4)*, we mean the convergent infinite series

$$R_n(x) = u_{n+1}(x) + u_{n+2}(x) + \ldots$$

We shall say that (4) converges *uniformly* on E·if (4) converges on E and if, for every $\varepsilon > 0$, an N exists such that, for $n > N$,

$$|R_n(x)| < \varepsilon$$

for every x of E.

If the sum of (4) is $s(x)$, then

$$R_n(x) = s(x) - s_n(x).$$

Hence, for (4) to converge uniformly on E, it is necessary and sufficient for the associated sequence

$$s_1(x), s_2(x), \ldots, s_n(x), \ldots$$

to converge uniformly on E.

The Weierstrass M-Test

3. *Theorem: Let*

$$M_1 + M_2 + \ldots + M_n + \ldots$$

be a convergent series of positive numbers. Let the series

(5) $$u_1(x) + u_2(x) + \ldots + u_n(x) + \ldots$$

be such that, for every x on E, $|u_n(x)| \leq M_n$. Then (5) converges absolutely and uniformly on E.

Proof: Let $\varepsilon > 0$ be given. Let N be such that for $n > N$ the remainder after n terms in the M series is less than ε. For any positive integer p, and for every x on E,

$$|u_{n+1}(x) + \ldots + u_{n+p}(x)| \leq M_{n+1} + \ldots + M_{n+p} < \varepsilon.$$

This shows that (5) converges throughout E. The remainder after n terms of (5) does not exceed in absolute value, for any x of E, the remainder after n terms of the M series. This proves uniform convergence. That the convergence is absolute is evident.

Example: The series

$$1 + \frac{1}{2} + \frac{1}{4} + \ldots + \frac{1}{2^n} + \ldots$$

converges. Hence

$$1 + \frac{x}{2} + \frac{x^2}{4} + \ldots + \frac{x^{n-1}}{2^{n-1}} + \ldots$$

converges absolutely and uniformly for $|x| \leq 1$.

CONTINUITY, INTEGRABILITY, DIFFERENTIABILITY

4. The equivalence of the theories of infinite series and infinite sequences leads immediately to the following three theorems:

(A) *Let*

(6) $$u_1(x) + \ldots + u_n(x) + \ldots$$

be uniformly convergent on the closed interval (a, b). Let each $u_n(x)$ be continuous at a point x_0 of (a, b). Then the sum of (6) is continuous at x_0.

(B) *Let*

(7) $$u_1(x) + \ldots + u_n(x) + \ldots$$

converge uniformly to a sum s(x) on a closed interval (a, b). Let each $u_n(x)$ be integrable on (a, b). Then

(a) *s(x) is integrable on (a, b),*

(b) $\int_a^b u_1(x)\, dx + \ldots + \int_a^b u_n(x)\, dx + \ldots$ *converges,*

(c) $\int_a^b s(x)\, dx = \int_a^b u_1(x)\, dx + \ldots + \int_a^b u_n(x)\, dx + \ldots.$

(C) *Let*

(8) $$u_1(x) + \ldots + u_n(x) + \ldots$$

converge on the closed interval (a, b), each $u_n(x)$ being differentiable throughout (a, b). Let $u_1'(x) +\ldots+ u_n'(x) +\ldots$ converge uniformly on (a, b). Then

(a) $u_1(x) +\ldots+ u_n(x) +\ldots$ converges uniformly on (a, b),

(b) the sum s(x) of (8) is differentiable on (a, b),

(c) $\frac{ds}{dx} = u_1'(x) +\ldots+ u_n'(x) +\ldots$.

A FUNCTION DISCONTINUOUS AT A DENSE SET OF POINTS, BUT INTEGRABLE

5. Let $\varphi(x) = -1$ for $x \leq 0$. Let $\varphi(x) = 1$ for $x > 0$. Then $\varphi(x)$ is continuous for $x \neq 0$ and discontinuous at $x = 0$.

We see that, if a is any constant, $\varphi(x - a)$ is continuous for every value of x except $x = a$.

Since $\varphi(x)$ is bounded and has only one discontinuity, $\varphi(x)$ is integrable in every interval. The same is true of $\varphi(x - a)$ for every a.

We take the interval (0, 1). Let

$$a_1, a_2, \ldots, a_n, \ldots$$

be any *countable* set of distinct points, dense in (0, 1); for instance, the set of all rational points in (0, 1).

Consider the expression

$$s(x) = \frac{\varphi(x - a_1)}{2} + \frac{\varphi(x - a_2)}{2^2} +\ldots+ \frac{\varphi(x - a_n)}{2^n} +\ldots$$

The series in the second member converges uniformly in (0, 1). In short, as $|\varphi(x - a_n)| = 1$ for every x, the terms of the series are equal in absolute value to the corresponding term of

$$1 + \frac{1}{2} +\ldots+ \frac{1}{2^n} +\ldots$$

so that the M-test applies.

Hence s(x) is integrable in (0, 1).

We shall show that s(x) is discontinuous at every point a_n.

The infinite series

$$s(x) - \frac{\varphi(x - a_n)}{2^n} = \frac{\varphi(x - a_1)}{2} +\ldots+ \frac{\varphi(x - a_{n-1})}{2^{n-1}} + \frac{\varphi(x - a_{n+1})}{2^{n+1}} +\ldots$$

converges uniformly on (0, 1). Its terms are all continuous at a_n. Hence $s(x) - \frac{\varphi(x - a_n)}{2^n}$ is continuous at a_n. But $\frac{\varphi(x - a_n)}{2^n}$ has a discontinuity at a_n. Thus s(x) is discontinuous at a_n.

X
Functions of Two Variables

POINTS

1. We consider, in what follows, pairs of real numbers (x, y). We use, as the geometric representation of such a number pair, a point of a plane with abscissa x and ordinate y. The plane and its points are introduced only to facilitate thought and expression and to assist the memory. Our reasoning will be purely arithmetic.

For the present *point* will mean a pair of real numbers. Any set of pairs of real numbers will be called a *two-dimensional point set*.

FUNCTIONS OF TWO VARIABLES

2. Consider any two-dimensional point set E. Suppose that with every (x, y) of E there is associated a real number z. We call z a *function of x and y*. We call E the *domain* of the *varibles* x and y. We write $z = f(x, y)$.

Examples:

(A) $$z = x + y$$

is defined for all real values of x and y.

(B) $$z = \frac{1}{x + y}$$

is defined for all points (x, y) for which $x + y \neq 0$.

RECTANGLES

3. By a *rectangle*, we shall mean a two-dimensional point set consisting of all points (x, y) for which

$$a < x < b$$
$$c < y < d$$

where a, b, c, d are real numbers with $b > a$, $d > c$.

For the present, we shall limit ourselves to functions defined in rectangles.

CONTINUITY

4. Let $f(x, y)$ be defined on a rectangle. Let (x_0, y_0) be any point of the rectangle. We shall say that $f(x, y)$ *is continuous in both variables at* (x_0, y_0) if, for every $\varepsilon > 0$ a $\delta > 0$ exists such that

$$|f(x, y) - f(x_0, y_0)| < \varepsilon$$

for $|x - x_0| < \delta$, $|y - y_0| < \delta$.

If x is held fast at x_0 and y is allowed to vary, $f(x, y)$ becomes a function of y alone, which we denote by $f(x_0, y)$. We call $f(x, y)$ *continuous in y at* (x_0, y_0) if $f(x_0, y)$ is continuous at y_0. Similarly, we define continuity in x alone.

A function can be continuous in x and y separately at a point without being continuous in both variables.

Example: Let $f(x, y) = 0$ when either x or y is zero. Let $f(x, y) = 1$ when neither x nor y is zero. Then $f(x, y)$ is continuous in x and y separately at $(0, 0)$. but is not continuous in both variables at $(0, 0)$.

PARTIAL DERIVATIVES

5. Let $f(x, y)$ be defined for all points of a rectangle. Let (x_0, y_0) be any point of the rectangle. Let h be any real number, distinct from 0, such that $(x_0 + h, y_0)$ lies within the rectangle. Consider the ratio

$$\frac{f(x_0 + h, y_0) - f(x_0, y_0)}{h}.$$

If as h approaches zero, this ratio approaches a limit, we call the limit the *partial derivative of $f(x, y)$ with respect to x at (x_0, y_0)*. In other words, $f(x, y)$ has a partial derivative with respect to x at (x_0, y_0) if $f(x, y_0)$ has a derivative with respect to x, in the ordinary sense at x_0. Formally, $f(x, y)$ has a partial derivative with respect to x at (x_0, y_0) if a number λ exists such that for every $\varepsilon > 0$, a $\delta > 0$ can be found such that

$$\left| \frac{f(x_0 + h, y_0) - f(x_0, y_0)}{h} - \lambda \right| < \varepsilon$$

for $0 < |h| < \delta$. The partial derivative will be λ.

If $f(x, y)$ has a partial derivative with respect to x at (x_0, y_0) we represent the partial derivative by

$$\left. \frac{\partial f}{\partial x} \right]_{\substack{x = x_0 \\ y = y_0}}$$

or by $\frac{\partial f(x_0, y_0)}{\partial x}$ or by $f_x(x_0, y_0)$.

Similarly, we define a partial derivative with respect to y, which we represent by

$$\left. \frac{\partial f}{\partial y} \right]_{\substack{x = x_0 \\ y = y_0}} \quad \text{or} \quad \frac{\partial f(x_0, y_0)}{\partial y} \quad \text{or} \quad f_y(x_0, y_0).$$

If $f(x, y)$ has a partial derivative with respect to x for each point of a rectangle, the partial derivative is a function of x and y defined in the rectangle. We represent this function by $\frac{\partial f}{\partial x}$ or f_x. Similarly, we define $\frac{\partial f}{\partial y}$.

If $f(x, y)$ has a partial derivative with respect to one of its variables, at a point, it is continuous in that variable at the point. but it is possible for $\frac{\partial f}{\partial x}$ and $\frac{\partial f}{\partial y}$ to exist at a point without f being continuous in both variables at the point. The example of §4 shows this.

Suppose that $\frac{\partial f}{\partial x}$ exists throughout a rectangle. It may be that $\frac{\partial f}{\partial x}$ has a partial derivative with respect to x at certain points. If so, we represent that derivative by

$$\frac{\partial^2 f}{\partial x^2}$$

Similarly, if $\frac{\partial f}{\partial x}$ has a partial derivative with respect to y, we represent that derivative by

$$\frac{\partial^2 f}{\partial y \, \partial x}.$$

In the same way, we represent the partial derivatives of $\frac{\partial f}{\partial y}$ with respect to x and y, when they exist, by

$$\frac{\partial^2 f}{\partial x\, \partial y} \text{ and } \frac{\partial^2 f}{\partial y^2}$$

respectively.

6. *Theorem: Let $f(x, y)$ be defined in a rectangle. Let $\frac{\partial^2 f}{\partial y\, \partial x}$ and $\frac{\partial^2 f}{\partial x\, \partial y}$ exist throughout the rectangle, and be continuous at some point (x_0, y_0) of the rectangle. Then*

$$\left.\frac{\partial^2 f}{\partial y\, \partial x}\right]_{\substack{x = x_0 \\ y = y_0}} = \left.\frac{\partial^2 f}{\partial x\, \partial y}\right]_{\substack{x = x_0 \\ y = y_0}}.$$

Remark: The hypothesis implies the existence of $\frac{\partial f}{\partial x}$ and $\frac{\partial f}{\partial y}$ throughout the rectangle.

Proof: Let h and k, distinct from zero, be such that

$$(x_0 + h, y_0),\ (x_0, y_0 + k),\ (x_0 + h, y_0 + k)$$

lie within the rectangle. We have, identically,

(1) $[f(x_0 + h, y_0 + k) - f(x_0, y_0 + k)] - [f(x_0 + h, y_0) - f(x_0, y_0)]$
$= [f(x_0 + h, y_0 + k) - f(x_0 + h, y_0)] - [f(x_0, y_0 + k) - f(x_0, y_0)].$

The first member of this equation is the difference between the two values of the function of y

$$f(x_0 + h, y) - f(x_0, y)$$

for the points $y_0 + k$ and y_0. This function of y has a derivative with respect to y throughout the interval $(y_0, y_0 + k)$. Hence, by the mean value theorem, the first member of (1) equals

(2) $\quad k\left[\frac{\partial f(x_0 + h, y_1)}{\partial y} - \frac{\partial f(x_0, y_1)}{\partial y}\right]$

where y_1 lies between y_0 and $y_0 + k$.

The quantity in brackets in (2) is the difference between the two values of the function of x

$$\frac{\partial f(x, y_1)}{\partial y}$$

for $x = x_0$ and $x = x_0 + h$. Then, by the mean value theorem, the quantity (2) equals

$$kh\, \frac{\partial^2 f(x_1, y_1)}{\partial x\, \partial y}$$

where x_1 lies between x_0 and $x_0 + h$.

Similarly, the second member of (1) is found to equal

$$hk\, \frac{\partial^2 f(x_2, y_2)}{\partial y\, \partial x}$$

where $x_0 < x_2 < x_0 + h;\ y_0 < y_2 < y_0 + k$.

Hence

$$kh\, \frac{\partial^2 f(x_1, y_1)}{\partial x\, \partial y} = hk\, \frac{\partial^2 f(x_2, y_2)}{\partial y\, \partial x},$$

so that, as $h \neq 0,\ k \neq 0$,

(3) $\quad\dfrac{\partial^2 f(x_1, y_1)}{\partial x\, \partial y} = \dfrac{\partial^2 f(x_2, y_2)}{\partial y\, \partial x}$

Now, when h and k are small, (x_1, y_1) and (x_2, y_2) will be very close to (x_0, y_0). Hence the two members of (3) will be very close to $\frac{\partial^2 f(x_0, y_0)}{\partial x \partial y}$ and $\frac{\partial^2 f(x_0, y_0)}{\partial y \partial x}$ respectively, because of the continuity assumed for the second derivatives. Hence we must have

$$\frac{\partial^2 f(x_0, y_0)}{\partial x \partial y} = \frac{\partial^2 f(x_0, y_0)}{\partial y \partial x}$$

else (3) could not hold for h and k small.

THE COMPLETE DIFFERENTIAL

7. Let $z = f(x, y)$ have continuous first partial derivatives throughout a rectangle. (This means that the derivatives are to be continuous in both variables.) Consider any point (x, y) of the rectangle. Let Δx and Δy be numbers such that $(x + \Delta x, y + \Delta y)$ lies in the rectangle. Let Δz represent

$$f(x + \Delta x, y + \Delta y) - f(x, y).$$

We shall prove that

$$\Delta z = \frac{\partial z}{\partial x} \Delta x + \frac{\partial z}{\partial y} \Delta y + \varepsilon \Delta x + \eta \Delta y$$

where ε and η are functions of Δx and Δy which, for any (x, y), approach 0 as Δx and Δy approach zero.

Proof: We have

$$\Delta z = [f(x + \Delta x, y + \Delta y) - f(x, y + \Delta y)] + [f(x, y + \Delta y) - f(x, y)],$$

$$= \frac{\partial f(x_1, y + \Delta y)}{\partial x} \Delta x + \frac{\partial f(x, y_1)}{\partial y} \Delta y$$

where x_1 lies between x and $x + \Delta x$ and y_1 between y and $y + \Delta y$.

Then

$$\Delta z = \frac{\partial f(x, y)}{\partial x} \Delta x + \frac{\partial f(x, y)}{\partial y} \Delta y + \left[\frac{\partial f(x_1, y + \Delta y)}{\partial x} - \frac{\partial f(x, y)}{\partial x}\right] \Delta x$$

$$+ \left[\frac{\partial f(x, y_1)}{\partial y} - \frac{\partial f(x, y)}{\partial y}\right] \Delta y.$$

Let

$$\varepsilon = \frac{\partial f(x_1, y + \Delta y)}{\partial x} - \frac{\partial f(x, y)}{\partial x}; \quad \eta = \frac{\partial f(x, y_1)}{\partial y} - \frac{\partial f(x, y)}{\partial y}$$

From the continuity of $\frac{\partial f}{\partial x}$ and $\frac{\partial f}{\partial y}$, we see that ε and η are small when Δx and Δy are small; that is, for every $h > 0$, a $k > 0$ exists such that $|\varepsilon| < h$, $|\eta| < h$, if $|\Delta x| < k$ and $|\Delta y| < k$.

XI

Complex and Hypercomplex Numbers

COMPLEX NUMBERS

1. By a *complex number*, we shall mean any *ordered pair of real numbers* (a, b). When we say that the number pairs are ordered, we mean that (a, b) and (b, a) are to be regarded as distinct if $a \neq b$.

ADDITION AND MULTIPLICATION

2. By $(a, b) + (c, d)$, we mean the complex number
$$(a + c, b + d).$$
It is seen at once that addition is associative and commutative.

3. By $(a, b) \times (c, d)$, we mean the complex number
$$(ac - bd, ad + bc).$$
It is seen immediately that multiplication, so defined, is commutative. Let us verify the associativity of multiplication. We have
$$(a, b)\,[(c, d)\,(e, f)] = (a, b)\,[ce - df, de + cf]$$
$$= (ace - adf - bde - bcf,\ ade + acf + bce - bdf).$$
Also
$$[(a, b)\,(c, d)]\,(e, f) = (ac - bd, ad + bc)\,(e, f)$$
$$= (ace - bde - adf - bcf,\ acf - bdf + ade + bce).$$
This proves the associativity. Let us prove that multiplication is distributive with respect to addition. We have
$$(a, b)\,[(c, d) + (e, f)] = (a, b)\,(c + e, d + f)$$
$$= (ac + ae - bd - bf,\ bc + be + ad + af)$$
$$= (ac - bd, bc + ad) + (ae - bf, be + af)$$
$$= (a, b)\,(c, d) + (a, b)\,(e, f).$$

Q.E.D.

SUBTRACTION AND DIVISION

4. It is evident that subtraction is always possible and that the result of subtraction is unique.

We shall prove that division by any complex number except $(0, 0)$ is possible and that the result is unique.

Given any (a, b), and any (c, d) with c and d not both zero, we seek an x and a y such that $(c, d)\,(x, y) = (a, b)$. We must have
$$cx - dy = a, \qquad dx + cy = b.$$

These two equations, if c and d are not both zero, have a unique solution

$$x = \frac{ac + bd}{c^2 + d^2}, \qquad y = \frac{bc - ad}{c^2 + d^2}.$$

This proves our statement.

If $(c, d) = (0, 0)$, there exists no (x, y) unless $(a, b) = (0, 0)$. When $(a, b) = (0, 0)$ and $(c, d) = (0, 0)$, we can take any complex number for (x, y).

CHANGE OF NOTATION

5. For complex numbers of the type $(a, 0)$, it will be convenient to use the simpler symbol a. Thus, a will sometimes represent the complex number $(a, 0)$ and sometimes the real number a. No confusion will arise.

No inconsistency can come out of this convention, since

$$(a, 0) + (b, 0) = (a + b, 0)$$
$$(a, 0) \times (b, 0) = (ab, 0).$$

We have, as may be verified by calculation,

$$(a, b) = (a, 0) + (b, 0)(0, 1).$$

For the complex number $(0, 1)$, we shall use the symbol i. We have thus, on the basis of what was said above,

$$(a, b) = a + bi.$$

Here, to repeat, in the second member, a means $(a, 0)$ and b means $(b, 0)$.

We have

$$i^2 = (0, 1)(0, 1) = (-1, 0)$$

so that $i^2 = -1$. Accordingly, we sometimes write $i = \sqrt{-1}$ and $(a, b) = a + b\sqrt{-1}$. Thus, $\sqrt{-1}$ is merely a new symbol for $(0, 1)$.

THE MODULUS

6. By the modulus of $a + bi$, we mean the non-negative square root of $a^2 + b^2$. We represent the modulus of $a + bi$ by $|a + bi|$. We have

$$|a + bi| = \sqrt{a^2 + b^2}.$$

In the second member, a and b are the real numbers a and b.

Theorem: $\qquad |(a + bi)(c + di)| = |a + bi|\,|c + di|.$

Proof: $\qquad |(a + bi)(c + di)| = |(ac - bd, ad + bc)|$
$$= \sqrt{(ac - bd)^2 + (ad + bc)^2}$$
$$= \sqrt{a^2c^2 + b^2d^2 + a^2d^2 + b^2c^2}$$
$$= \sqrt{a^2 + b^2} \cdot \sqrt{c^2 + d^2}$$
$$= |a + bi| \cdot |c + di|.$$

GEOMETRIC REPRESENTATION

7. To represent the point $a + bi$, we use, in a plane, a point of abscissa a and ordinate b. Without corrupting our arithmetic rigor, we mention that the geometric counterpart of the modulus is the distance from the plotted point to the origin.

MODULUS OF SUM AND DIFFERENCE

8. *Theorem: If α and β are two complex numbers*
$$|\alpha + \beta| \leq |\alpha| + |\beta|.$$

Proof: Let $\alpha = a + bi$, $\beta = c + di$, so that $\alpha + \beta = (a + c) + (b + d)i$.
Then
$$|\alpha + \beta|^2 = (a + c)^2 + (b + d)^2$$
and
$$(|\alpha| + |\beta|)^2 = |\alpha|^2 + |\beta|^2 + 2|\alpha||\beta|$$
$$= a^2 + b^2 + c^2 + d^2 + 2\sqrt{(a^2 + b^2)(c^2 + d^2)}.$$

Hence
$$|\alpha + \beta|^2 - (|\alpha| + |\beta|)^2 = 2[ac + bd - \sqrt{(a^2 + b^2)(c^2 + d^2)}].$$

Now
$$(a^2 + b^2)(c^2 + d^2) = (ac + bd)^2 + (ad - bc)^2.$$

Hence
$$\sqrt{(a^2 + b^2)(c^2 + d^2)} \geq |ac + bd|.$$

Hence
$$|\alpha + \beta|^2 - (|\alpha| + |\beta|)^2 \leq 0$$
$$|\alpha + \beta|^2 \leq (|\alpha| + |\beta|)^2$$
$$|\alpha + \beta| \leq |\alpha| + |\beta|.$$
Q.E.D.

Theorem: $\quad |\alpha - \beta| \geq |\alpha| - |\beta|.$

Proof:
$$\alpha = \beta + (\alpha - \beta),$$
$$|\alpha| \leq |\beta| + |\alpha - \beta|,$$
$$|\alpha - \beta| \geq |\alpha| - |\beta|.$$
Q.E.D.

HYPERCOMPLEX NUMBERS AND LINEAR ALGEBRAS

9. Let n be any positive integer. We shall call the totality of ordered sets of n real numbers a *linear algebra* under the following circumstances.

There must be defined, for the sets of n numbers, two operations, called addition and multiplication. The definition of addition must be
$$(a_1, \ldots, a_n) + (b_1, \ldots, b_n) = (a_1 + b_1, \ldots, a_n + b_n).$$

The definition of multiplication must be of the following type. There must exist n^3 numbers
$$\gamma_{ijk}, \; i, j, k = 1, \ldots, n,$$
such that, for any two two sets $(a_1, \ldots, a_n)$ and $(b_1, \ldots, b_n)$, we have
$$(a_1, \ldots, a_n)(b_1, \ldots, b_n) = (c_1, \ldots, c_n),$$
where
$$c_i = \Sigma \gamma_{ijk} a_j b_k, \; j, k = 1, \ldots, n.$$

The sets $(a_1, \ldots, a_n)$ are called hypercomplex numbers.

XII

Limits and Point Sets (Complex Domain)

LIMITS

1. Consider a sequence of complex numbers

$$(1) \qquad z_1, z_2, \ldots, z_n, \ldots,$$

where $z_n = x_n + iy_n$.

The sequence (1) is said to *converge* if a complex number λ exists such that, for every positive real ϵ, an N can be found such that, for $n > N$,

$$|\lambda - z_n| < \epsilon.$$

Such a number λ, if it exists, is said to be a *limit* of the sequence (1). It will be seen below that a sequence of complex numbers cannot have more than one limit.

Using the representation $z_n = x_n + iy_n$, let us form the sequences of real numbers

$$(2) \qquad x_1, x_2, \ldots, x_n, \ldots,$$
$$(3) \qquad y_1, y_2, \ldots, y_n, \ldots.$$

In (2) and (3), x_n and y_n are supposed to be real numbers, rather than the number pairs $(x_n, 0)$, $y_n, 0)$.

We say that, for $\lambda = u + iv$ to be a limit of (1), it is necessary and sufficient that (2) have u as a limit and that (3) have v as a limit.

Proof: (a) *Necessity:* Let $u + iv$ be a limit of (1). Taking any $\epsilon > 0$, suppose that $|(u + iv) - (x_n + iy_n)| < \epsilon$ for $n > N$.

Now

$$|(u + iv) - (x_n + iy_n)| = |(u - x_n) + i(v - y_n)|$$
$$= \sqrt{(u - x_n)^2 + (v - y)^2}.$$

Hence

$$|(u + iv) - (x_n + iy_n)| \geq |u - x_n|$$

and

$$|(u + iv) - (x_n + iy_n)| \geq |v - y_n|.$$

Then $|u - x_n| < \epsilon$ and $|v - y_n| < \epsilon$ for $n > N$, which shows that (2) converges to u and (3) to v.

Q.E.D.

(b) *Sufficiency:* Let (2) converge to u and (3) to v. Given an $\epsilon > 0$, take N so that $|u - x_n| < \epsilon$ and $|v - y_n| < \epsilon$ for $n > N$. Then, for $n < N$,

$$|(u + iv) - (x_n + iy_n)| = \sqrt{(u - x_n)^2 + (v - y_n)^2}$$
$$< \sqrt{\epsilon^2 + \epsilon^2}$$
$$= \epsilon\sqrt{2}.$$

Q.E.D.

As (2) and (3) cannot have more than one limit, we see that (1) cannot have more than one limit.

OPERATIONS WITH SEQUENCES

2. *Theorem:* Let

(4) $\quad z_1, z_2, \ldots, z_n, \ldots$

(5) $\quad w_1, w_2, \ldots, w_n, \ldots$

be two convergent sequences of complex numbers. Then

(6) $\quad z_1 + w_1, z_2 + w_2, \ldots, z_n + w_n, \ldots$

is convergent and its limit is the sum of the limits of (4) and (5).

Proof: Let α and β be respectively the limits of (4) and (5). Given an $\epsilon > 0$, take N such that $|\alpha - z_n| < \epsilon$, $|\beta - w_n| < \epsilon$ for $n > N$. Then, for $n > N$,

$$|(\alpha + \beta) - (z_n + w_n)| = |(\alpha - z_n) + (\beta - w_n)| \leq |\alpha - z_n| + |\beta - w_n| < \epsilon + \epsilon = 2\epsilon.$$

Q.E.D.

Theorem: Let

(7) $\quad z_1, z_2, \ldots, z_n, \ldots$

(8) $\quad w_1, w_2, \ldots, w_n, \ldots$

be two convergent sequences of complex numbers. Then

$$z_1 w_1, z_2 w_2, \ldots, z_n w_n, \ldots$$

is convergent and its limit is the product of the limits of (7) and of (8).

Proof: As for real sequences.

Theorem: Let

(9) $\quad z_1, z_2, \ldots, z_n, \ldots$

converge and let

(10) $\quad w_1, w_2, \ldots, w_n, \ldots$

where no w_n is zero, converge to a limit distinct from zero. Then

$$\frac{z_1}{w_1}, \ldots, \frac{z_n}{w_n}, \ldots$$

converges to a limit which is the quotient of the limits of (9) and (10).

Proof: As for real sequences.

POINT SETS IN THE COMPLEX DOMAIN

3. By a *point*, we shall mean, in what follows, a complex number.

Any aggregate of points will be called a *point set*.

Let z_0 be any point. By the *interior of a circle with z_0 as center*, we shall mean the set of points z such that

$$|z - z_0| < \epsilon,$$

where ϵ is any positive real number.

By a *neighborhood* of a point z_0, we shall mean the interior of any circle with z_0 as center. Every point z_0 has an infinite number of neighborhoods.

Consider any point set E. A point z will be called a limit point of E if every neighborhood of z contains a point of E distinct from z. If z is a limit point of E, every neighborhood of z contains an infinite number of points of E.

Examples: If E is the whole plane, every point z is a limit point of E. If E is the set of all points $x + iy$ with x and y rational, then every point z is a limit point of E. A convergent sequence of distinct points has one and only one limit point, namely, the limit of the sequence.

If E is a *rectangle*, that is the totality of those points $x + yi$ for which

$$a < x < b$$
$$c < y < d$$

where $b > a$ and $d > c$, then every point of the *closed rectangle*

$$a \leq x \leq b$$
$$c \leq y \leq d$$

is a limit point of E.

A point set E will be called closed if either:

(a) *E has limit points and each of the limit points of E belongs to E*

or

(b) *E has no limit points.*

Examples under (a): The whole plane (that is, the set of all complex numbers) is closed. What we have called above a *closed rectangle* is closed. The set of points consisting of a convergent sequence of points and of the limit of the sequence, is closed.

The set of all numbers $x + yi$ with x and y rational is not closed.

A set E is called *perfect* if it is closed and if every point of E is a limit point of E.

THE THEOREM ON NESTED RECTANGLES

4. Let $R_1, R_2, \ldots, R_n, \ldots$ be a sequence of closed rectangles, each of which contains its successor. That is, each R_n is a set of points $x + iy$ for which

$$a_n \leq x \leq b_n$$
$$c_n \leq y \leq d_n$$

where $b_n > a_n$, $d_n > c_n$, and each point of R_n with $n > 1$ is contained in R_{n-1}.

Suppose that the dimensions of the rectangles R_n tend towards zero, that is, that the sequences

$$b_1 - a_1, b_2 - a_2, \ldots, b_n - a_n, \ldots,$$
$$d_1 - c_1, d_2 - c_2, \ldots, d_n - c_n, \ldots$$

both converge to zero.

We say that there is one and only one point which is contained in every rectangle R_n.

Proof: (a) *Existence.* Each of the closed intervals

$$(a_1, b_1), \ldots, (a_n, b_n), \ldots$$

contains its successor and the lengths of the intervals converge to zero. Hence there is a point u (a real number) common to all of them. Similarly, there is a point v common to all intervals (c_n, d_n).

Then u + iv is contained in each rectangle R_n.

(b) *Uniqueness.* If u + iv is contained in each R_n then u lies in every interval (a_n, b_n) and v in every interval (c_n, d_n). Hence there is but one possibility for u and one possibility for v.

BOREL'S THEOREM

5. A point set E is said to be *bounded* if there exists a rectangle which contains every point of E.

Theorem: Let E be a closed and bounded point set. Suppose that there is given an infinite set of circle interiors such that each point of E is contained in at least one of the circle interiors. Then there is some finite subset of the given infinite set of circle interiors such that each point of E is contained in at least one circle interior of the finite subset.

Proof: Suppose that there exists no finite subset of interiors of circles as described above.

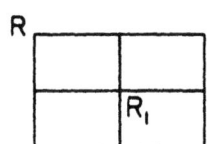

Let R be a closed rectangle which contains E. Divide R into four equal closed rectangles.

There must be one of these four closed rectangles, say R_1, such that the points of E lying in R_1 cannot be covered by a finite number of the given circle interiors. Divide R_1 into four equal rectangles and continue.

We form thus an infinite sequence of closed rectangles

$$R_1, R_2, R_3, \ldots$$

each containing its successor, the dimensions of the rectangles tending towards zero.

There exists a point P which is contained in every R_n.

Now P is a limit point of E, for any neighborhood of P will contain R_n if n is sufficiently large and every R_n contains an infinite number of points of E.

Because E is closed, P is contained in E.

Then there exists a circle interior of the given set which contains P. Such a circle interior will contain R_n if n is large. This contradicts the deduction that the points of E in R_n cannot be covered by a finite number of the given circle interiors.

Q.E.D.

XIII

Curves and Regions

INTERIOR, EXTERIOR AND BOUNDARY POINTS

1. Consider any point set E in the complex plane. We separate the points of the complex plane into the following three classes relative to E:

(a) Interior points;
(b) Exterior points;
(c) Boundary points.

(a) A point z will be said to be *interior to* E if there exists a neighborhood of z every point of which belongs to E.

(b) A point z will be said to be *exterior to* E if there exists a neighborhood of z no point of which belongs to E.

(c) A point z will be called a *boundary point of* E if every neighborhood of z contains points belonging to E and points not belonging to E.

Of course, a point interior to E belongs to E and a point exterior to E does not belong to E. A boundary point of C may either belong to E or not belong to E.

Examples:

If E is the entire plane, every point of the plane is interior to E.

If E is the circle interior consisting of all points z such that

$$|z| < a$$

where $a > 0$, every z with $|z| < a$ is interior to E, every z with $|z| > a$ is exterior to E and every z with $|z| = a$ is a boundary point of E.

If E is the point 0, every point except 0 is exterior to E and 0 is a boundary point of E.

If E consists of all real points, that is, all points of the type $a + 0i$, every point not in E is exterior to E and the real points are boundary points of E.

If E consists of all points $x + yi$ with x and y both rational, then every point of the plane is a boundary point of E.

CONTINUOUS CURVES

2. Let

$$x = \varphi(t)$$

and

$$y = \psi(t)$$

be two real functions of the real variable t, continuous on the closed interval (a, b). We shall call such a pair of functions a *continuous curve*.

Consider, for any particular t on (a, b), the point of the complex plane

$$x + yi = \varphi(t) + i\psi(t).$$

The totality of such points obtained by allowing t to range over the closed interval (a, b) will be called the *points* of the curve

$$x = \varphi(t), \; y = \psi(t); \; a \leq t \leq b.$$

Examples:

(a) Let

$$x = \alpha t + \beta, \; y = \gamma t + \delta$$

with α, β, γ, δ real numbers. For t on an interval (a, b), these equations define a continuous curve which is called a *segment of a straight line*.

(b) Let

$$x = t$$
$$y = \sqrt{1 - t^2} \text{ (non-negative square root)},$$

where the interval is (-1, 1). The curve is *the upper half of a circle*.

The curve $x = \varphi(t)$, $y = \psi(t)$, $a \leq t \leq b$ will be said to join the points z_1 and z_2 if

$$z_1 = \varphi(a) + i\psi(a), \; z_2 = \varphi(b) + i\psi(b).$$

OPEN REGIONS

3. A point set E will be called an *open region* if the following two conditions are satisfied:

(a) Every point of E is interior to E.

(b) Any two points of E can be joined by a continuous curve all of whose points belong to E.

FUNCTIONS AND CONTINUITY

DEFINITION OF FUNCTION

1. Consider any set of points E of the complex plane. Suppose that, with every z of E, there is associated a complex number w. We call w a *function of z*. The set E is called the domain of z. The function w is said to be *defined on E*.

Examples: The function $w = z^2$ is defined for the whole plane. The function $w = \dfrac{1}{z}$ is defined for every value of z except $z = 0$.

CONTINUITY

2. Let

$$w = f(z)$$

be defined on a set E which has limit points, some of the limit points of E belonging to E. Let z_0 be a limit point of E which belongs to E. Suppose that, for every $\varepsilon > 0$, we can find a $\delta > 0$ such that, if $|z - z_0| < \delta$, where z is in E, we have

$$|f(z) - f(z_0)| < \varepsilon.$$

In that case, we shall say that $f(z)$ is *continuous at* z_0.

Thus, continuity is a notion which can be applied only to points of E which are limit points of E.

UNIFORM CONTINUITY

3. Let $w = f(z)$ be defined on a set E which is such that every point of E is a limit point

of E. Suppose that, for every $\varepsilon > 0$, we can find a $\delta > 0$ such that, if $|z_2 - z_1| < \delta$, where z_1 and z_2 belong to E, we have

$$|f(z_2) - f(z_1)| < \varepsilon.$$

In that case, we shall say that $f(z)$ is *uniformly continuous* on E.

Theorem: Let $w = f(z)$ *be continuous at each point of a bounded perfect set. Then $f(z)$ is uniformly continuous on the set.*

Proof: Let E designate the bounded perfect set. Let an $\varepsilon > 0$ be assigned. For every z of E there is a $\delta_z > 0$ such that, if $|z' - z| < \delta_z$, with z' in E, we have

$$|f(z') - f(z)| < \frac{\varepsilon}{2}.$$

Let each point z of E be enclosed in a circle interior with center at z and radius $\frac{1}{2}\delta_z$. Since E is *closed and bounded*, there exist a finite number of these circle interiors which contain all of the points of E. Let δ be the least of the radii of such a finite number of circles.

We say that, if z_2 and z_1 belong to E and if $|z_2 - z_1| < \delta$, then $|f(z_2) - f(z_1)| < \varepsilon$. This will mean, of course, that $f(z)$ is uniformly continuous on E.

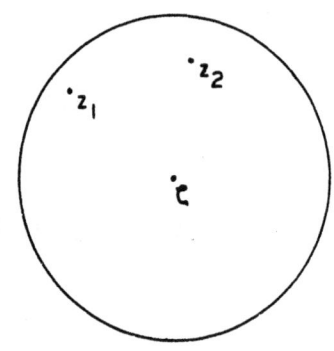

Let ζ be the center of any circle of the finite set considered above which contains z_1. The radius of this circle is $\frac{1}{2}\delta_\zeta$, which is at least as great as δ.

As

$$|z_1 - \zeta| < \frac{1}{2}\delta_\zeta$$

and as

$$|z_2 - z_1| < \delta \leq \frac{1}{2}\delta_\zeta,$$

we have

$$|z_2 - \zeta| = |(z_2 - z_1) + (z_1 - \zeta)| < \delta_\zeta.$$

Hence

$$|f(z_1) - f(\zeta)| < \frac{\varepsilon}{2}$$

and

$$|f(z_2) - f(\zeta)| < \frac{\varepsilon}{2}.$$

Thus,

$$|f(z_2) - f(z_1)| < \varepsilon.$$

Q.E.D.

XIV

Derivatives

DERIVATIVE

1. Let
$$w = f(z)$$
be defined for every point of an open region.

Let z_0 be some point of the open region. If $|h|$ is sufficiently small, then $z_0 + h$ will belong to the open region.

Suppose that, as h approaches zero,
$$\frac{f(z_0 + h) - f(z_0)}{h}$$
tends toward a limit; that is, suppose that a complex number λ exists such that, for every $\varepsilon > 0$, a $\delta > 0$ exists such that
$$\left|\frac{f(z_0 + h) - f(z_0)}{h} - \lambda\right| < \varepsilon$$
for $|h| < \delta$. In that case, we shall call λ the *derivative* of $f(z)$ at z_0.

For instance, let $f(z) = z^2$. Then
$$\frac{f(z_0 + h) - f(z_0)}{h} = \frac{(z_0 + h)^2 - z_0^2}{h} = \frac{2hz_0 + h^2}{h} = 2z_0 + h.$$

It is thus plain that, for any z_0, the function z^2 has a derivative equal to $2z_0$.

Similarly the function
$$a_0 z^n + a_1 z^{n-1} + \ldots + a_n$$
with constant a_i, has, for every z, a derivative equal to
$$na_0 z^{n-1} + (n - 1) a_1 z^{n-2} + \ldots + a_{n-1}.$$

Evidently, if $f(z)$ has a derivative at z_0, $f(z)$ is continuous at z_0.

MONOGENICITY

2. If $f(z)$ has a derivative at z_0, we shall say that $f(z)$ is *monogenic* at z_0. This word is employed to emphasize the fact that
$$\frac{f(z_0 + h) - f(z_0)}{h}$$
approaches a limit λ for any sequence of values of h,
$$h_1, h_2, \ldots, h_n, \ldots$$
tending toward zero, λ being the same for all sequences of this type.

It means a good deal more for a function of a complex variable to have a derivative than for a function of a real variable to have a derivative. The reason for this is that, in the difference quotient for a function of a complex variable, h has two degrees of freedom; its real part and its pure imaginary part can be taken arbitrarily.

To obtain a function of a real variable which is continuous at a point and lacks a derivative at the point requires a little search. To construct a continuous function of a real variable which lacks a derivative everywhere, or even at a dense set of points, requires considerable effort.

On the other hand, most continuous functions of a complex variable which one might write down at random do not have derivatives, that is, they are not monogenic.

For example, let
$$w = f(z) = x - yi,$$
it being understood that $z = x + yi$. Here w is a function of z, because to know z is to know x and y and hence $x - yi$. We shall see whether this function has a derivative at $z = 0$. The difference quotient may be written as
$$\frac{f(z) - f(0)}{z - 0} = \frac{x - yi}{x + yi}.$$

Now
$$\frac{x - yi}{x + yi} = \frac{(x - yi)^2}{x^2 + y^2} = \frac{x^2 - y^2}{x^2 + y^2} - \frac{2xy}{x^2 + y^2} i.$$

Let m be any real number. Let z approach zero in such a way that y always equals mx. The difference quotient may be written
$$\frac{1 - m^2}{1 + m^2} - \frac{2m}{1 + m^2} i.$$

The limit of this expression, as z approaches zero, is the expression itself. It is thus plain that the limit of the difference quotient will be different for different methods of approach to the origin.

It is also possible to let z approach the origin in such a way that y/x does not approach a limit and then the difference quotient will usually not approach a limit at all.

Thus, $f(z)$ is not monogenic at $z = 0$. In the same way, one can prove that $f(z)$ is not monogenic for any value of z.

THE CAUCHY-RIEMANN EQUATIONS

3. We are going to derive conditions for a function
$$w = f(z)$$
to be monogenic at a point z. Let, for any value of z,
$$f(z) = u + iv$$
where u and v are real. Then u and v are functions of x and y. We write
$$f(z) = u(x, y) + iv(x, y).$$
We use Δz, rather than h, as a symbol for the increment of z. We write
$$\Delta z = \Delta x + i\Delta y.$$
Then
$$f(z + \Delta z) = u(x + \Delta x, y + \Delta y) + iv(x + \Delta x, y + \Delta y).$$
Consider some particular value of z. It is easy to see that if $f(z)$ is continuous for this value of z, then u and v are continuous in both variables for the corresponding x and y. For

$$|f(z + \Delta z) - f(z)| = |u(x + \Delta x, y + \Delta y) - u(x, y) + i[v(x + \Delta x, y + \Delta y) - v(x, y)]|$$
$$= \sqrt{[u(x + \Delta x, y + \Delta y) - u(x, y)]^2 + [v(x + \Delta x, y + \Delta y) - v(x, y)]^2}.$$

Thus
$$|f(z + \Delta z) - f(z)| \geq |u(x + \Delta x, y + \Delta y) - u(x, y)|,$$
and
$$|f(z + \Delta z) - f(z)| \geq |v(x + \Delta x, y + \Delta y) - v(x, y)|,$$
from which the continuity of u and v follows.

Suppose now that $f(z)$ has a derivative $f'(z)$ at the point z under consideration. In the difference quotient
$$\frac{f(z + \Delta z) - f(z)}{\Delta z}$$
we shall let Δz tend toward zero through real values; that is, in
$$\Delta z = \Delta x + i\Delta y,$$
we shall keep Δy equal to zero, so that $\Delta z = \Delta x$. Then
$$\frac{f(z + \Delta z) - f(z)}{\Delta z} = \frac{u(x + \Delta x, y) - u(x, y)}{\Delta x} + i \frac{v(x + \Delta x, y) - v(x, y)}{\Delta x}$$

When Δz approaches 0,
$$\frac{u(x + \Delta x, y) - u(x, y)}{\Delta x}$$
must approach the real part of $f'(z)$ and
$$\frac{v(x + \Delta x, y) - v(x, y)}{\Delta x}$$
must approach the coefficient of i in $f'(z)$. This means, firstly, that u and v have partial derivatives with respect to x at the point considered and, secondly, that

(1) $$f'(z) = \frac{\partial u}{\partial x} + i \frac{\partial v}{\partial x}.$$

Suppose now that Δz tends toward zero through pure imaginary values; that is, in
$$\Delta z = \Delta x + i\Delta y,$$
let $\Delta x = 0$, so that $\Delta z = i\Delta y$. Then
$$\frac{f(z + \Delta z) - f(z)}{\Delta z} = \frac{u(x, y + \Delta y) - u(x, y)}{i\Delta y} + i \frac{v(x, y + \Delta y) - v(x, y)}{i\Delta y},$$
$$= \frac{v(x, y + \Delta y) - v(x, y)}{\Delta y} - i \frac{u(x, y + \Delta y) - u(x, y)}{\Delta y}.$$

It follows that v and u have partial derivatives with respect to y at the point (x, y) considered and that

(2) $$f'(z) = \frac{\partial v}{\partial y} - i \frac{\partial u}{\partial y}.$$

Comparing (1) and (2), we have
$$\frac{\partial u}{\partial x} + i \frac{\partial v}{\partial x} = \frac{\partial v}{\partial y} - i \frac{\partial u}{\partial y}.$$

Then

$$\frac{\partial u}{\partial x} = \frac{\partial v}{\partial y},$$

$$\frac{\partial u}{\partial y} = -\frac{\partial v}{\partial x}.$$

These equations are called the *Cauchy-Riemann* equations. They are satisfied at every point at which $f(z)$ is monogenic.

The question arises as to whether the satisfaction of the Cauchy-Riemann equations at a point is sufficient for $f(z)$ to be monogenic at the point. We shall prove, in this connection, the following result.

Consider an open region in the complex plane. Let E represent the two-dimensional point set of points (x, y) obtained from the points z in the given open region. That is, if $z = x + yi$ is in the open region, then (x, y) is in E. Of course, the geometric picture of E is identical with that of the open region.

Let $u(x, y)$ and $v(x, y)$ be two real functions of x and y which have, throughout E, *continuous* first partial derivatives which satisfy the Cauchy-Riemann equations. Then the function of z given by

$$f(z) = u(x, y) + iv(x, y)$$

is monogénic at every point of the open region.

Proof: Let z be any point of the open region. Then

$$f(z + \Delta z) - f(z) = [u(x + \Delta x, y + \Delta y) - u(x, y)] + i[v(x + \Delta x, y + \Delta y) - v(x, y)].$$

Now, since u and v have *continuous* first partial derivatives, we have, for Δx and Δy small,

$$u(x + \Delta x, y + \Delta y) - u(x, y) = \frac{\partial u}{\partial x} \Delta x + \frac{\partial u}{\partial y} \Delta y + \varepsilon_1 \Delta x + \eta_1 \Delta y,$$

$$v(x + \Delta x, y + \Delta y) - v(x, y) = \frac{\partial v}{\partial x} \Delta x + \frac{\partial v}{\partial y} \Delta y + \varepsilon_2 \Delta x + \eta_2 \Delta y,$$

where $\varepsilon_1, \eta_1, \varepsilon_2, \eta_2$, approach zero as Δx and Δy approach zero.

Then

$$f(z + \Delta z) - f(z) = [\frac{\partial u}{\partial x} \Delta x + i \frac{\partial v}{\partial y} \Delta y] + [\frac{\partial u}{\partial y} \Delta y + i \frac{\partial v}{\partial x} \Delta x]$$

$$+ \Delta x(\varepsilon_1 + i\varepsilon_2) + \Delta y(\eta_1 + i\eta_2).$$

By the Cauchy-Riemann equations,

$$f(z + \Delta z) - f(z) = \frac{\partial u}{\partial x}(\Delta x + i\Delta y) + i\frac{\partial v}{\partial x}(\Delta x + i\Delta y) + \Delta x(\varepsilon_1 + i\varepsilon_2) + \Delta y(\eta_1 + i\eta_2).$$

Hence

$$\frac{f(z + \Delta z) - f(z)}{\Delta z} = \frac{\partial u}{\partial x} + i\frac{\partial v}{\partial x} + \frac{\Delta x}{\Delta x + i\Delta y}(\varepsilon_1 + i\varepsilon_2) + \frac{\Delta y}{\Delta x + i\Delta y}(\eta_1 + i\eta_2).$$

Now

$$\left|\frac{\Delta x}{\Delta x + i\Delta y}\right| \leq 1, \quad \left|\frac{\Delta y}{\Delta x + i\Delta y}\right| \leq 1.$$

It is thus plain that as Δz approaches zero,

$$\frac{\Delta x}{\Delta x + i\Delta y}(\varepsilon_1 + i\varepsilon_2) + \frac{\Delta y}{\Delta x + i\Delta y}(\eta_1 + i\eta_2)$$

approaches zero. Then, when Δz approaches zero,

$$\frac{f(z) + \Delta z) - f(z)}{\Delta z}$$

approaches a limit equal to $\frac{\partial u}{\partial x} + i\frac{\partial v}{\partial x}$

Q.E.D.

Note: In the statement of the theorem, it was assumed that the Cauchy-Riemann equations are satisfied throughout E. This led to the conclusion that $f(z)$ is monogenic at every point of the open region. But it is clear from the proof that the existence and continuity of the partial derivatives of u and v throughout E and the satisfaction of the C.-R. equations for a single value of z imply monogenicity for that value of z.

THE LAPLACE EQUATION

4. Let $u(x, y)$ and $v(x, y)$ be as in the preceding section and, in addition, let $u(x, y)$ and $v(x, y)$ have continuous second partial derivatives throughout E. From the C.-R. equations, which tell us that

$$\frac{\partial u}{\partial x} = \frac{\partial v}{\partial y}, \qquad \frac{\partial u}{\partial y} = -\frac{\partial v}{\partial x},$$

we find

$$\frac{\partial^2 u}{\partial x^2} = \frac{\partial^2 v}{\partial x \partial y}, \qquad \frac{\partial^2 u}{\partial y^2} = -\frac{\partial^2 v}{\partial y \partial x}$$

As $\frac{\partial^2 v}{\partial x \partial y}$ and $\frac{\partial^2 v}{\partial y \partial x}$ are continuous, they are equal. Then

$$\frac{\partial^2 u}{\partial x^2} + \frac{\partial^2 u}{\partial y^2} = 0$$

This equation is called *Laplace's Equation*. For v we have similarly,

$$\frac{\partial^2 v}{\partial x^2} + \frac{\partial^2 v}{\partial y^2} = 0.$$

A solution of Laplace's equation is called a *harmonic function*. The functions u and v above, considered as a pair, are called *conjugate harmonic functions*.

Note: The existence and continuity of the second derivatives of u and v, assumed above, is really a consequence of the assumption made for u and v in §3. This will be proved later.

DEFINITION OF ANALYTIC FUNCTION

5. A function of z which is defined in an open region and has a derivative at each point of the open region will be said to be *analytic* in the open region.

CONFORMAL MAPPING

6. Let $w = f(z)$ be analytic in an open region. For every z in the open region, there is a w. Consider any curve in the open region. The points w which correspond to the points of this curve will, if plotted in a w-plane, give a curve which may be regarded as an image (picture) of the curve in the z-plane. Consider two curves of the z-plane which intersect at a certain angle. (Note that we are speaking informally. The present material is a geometric application of function theory, rather than pure function theory.) Let z_0 be the point of intersection. We shall prove that, if $f'(z_0) \neq 0$, the images of the two curves in the w-plane intersect at the same angle.

Proof: We shall use freely the geometric theory of complex numbers which is presented in courses on algebra. This is legitimate, because the present discussion is parenthetical and will not be used at all in our later work.

By amp $(z_2 - z_1)$, (amplitude of $z_2 - z_1$), we shall mean the angle which the line joining z_1 and z_2 makes with the positive real axis. In particular, amp z is the angle between the positive real axis and the line which joins the origin to z. It is shown in texts on algebra that the amplitude of the quotient of two numbers is the difference of their amplitudes.

Let z_1 be a point of the first curve in the z-plane, distinct from z_0. As z_1 approaches z_0,
$$\text{amp } (z_1 - z_0)$$
will approach the inclination of the tangent to the curve at z_0.

Similarly,
$$\text{amp } (w_1 - w_0)$$
approaches the inclination of the tangent at w_0 to the image of the first curve.

Hence
$$\text{limit } [\text{amp } (w_1 - w_0) - \text{amp } (z_1 - z_0)]$$
which equals
$$\text{limit amp } \frac{w_1 - w_0}{z_0 - z_0}$$
is the difference between the inclination of the tangent to the first image curve at w_0 and the inclination of the tangent at z_0 to the first curve in the z-plane. Now, if $f'(z_0) \neq 0$,
$$\text{limit amp } \frac{w_1 - w_0}{z_1 - z_0} = \text{amp } f'(z_0).$$

Similarly, the difference between the inclinations of the second w-curve and the second z-curve at w_0 and z_0 respectively is amp $f'(z_0)$.

Let θ_1 and θ_2 be the inclinations of the tangents to the two z-curves at z_0 and ζ_1 and ζ_2 the inclinations of the tangents to the two w-curves.

Then
$$\zeta_1 - \theta_1 = \zeta_2 - \theta_2$$

Hence
$$\zeta_1 - \zeta_2 = \theta_1 - \theta_2$$

That is, the angle between the two w-curves equals the angle between the two z-curves.

Q.E.D.

XV

Continuous Curves

INVERSE FUNCTIONS (REAL VARIABLE)

1. Let $y = f(x)$ be continuous and increasing on the closed interval (a, b). Let

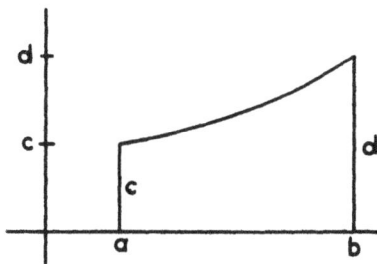

$$f(a) = c, \quad f(b) = d.$$

Then on (a, b), y assumes exactly once every value in the interval (c, d).

Hence to every value of y between c and d, there corresponds one value of x between a and b. We may thus regard x as a function of y, defined on (c, d). We write $x = g(y)$.

It is obvious that x is an increasing function of y on (c, d). We are going to prove that x is a continuous function of y on (c, d). Let y_0 be any value of y, and x_0 the corresponding value of x. Let x_0 be interior to (a, b). The cases of $x = a$ and $x = b$ are treated in the manner in which we treat values of x_0 which are interior points of (a, b).

Let a small positive number ε be assigned.

Let
$$f(x_0 - \varepsilon) = y_0 - \delta_1, \quad f(x_0 + \varepsilon) = y_0 + \delta_2.$$

Then δ_1 and δ_2 are positive. Let δ be the smaller of δ_1 and δ_2. Then if y lies between $y_0 - \delta$ and $y_0 + \delta$, x will lie between $x_0 - \varepsilon$ and $x_0 + \varepsilon$. This proves that $g(y)$ is continuous at y_0.

We shall call $g(y)$ the *function inverse* to $f(x)$.

A similar theorem holds for decreasing functions.

ON THE POINTS OF A CONTINUOUS CURVE

2. *Theorem:* If the points of a continuous curve consist of more than one point, they constitute a perfect set.

Proof: Let
$$x = \varphi(t), \quad y = \psi(t), \quad a \leq t \leq b$$
be a continuous curve.

We shall prove first that the points of the curve form a closed set.

Suppose that $P = \xi + i\eta$ is a limit point of points of the curve. Let
$$\varphi(t_1) + i\psi(t_1), \ldots, \quad \varphi(t_n) + i\psi(t_n), \ldots$$
be a sequence of distinct points of the curve approaching P. Then
$$t_1, t_2, \ldots, t_n, \ldots$$
are all distinct. By the Bolzano-Weierstrass theorem, the points of this sequence have at least one limit point. Of course, such a limit point lies in (a, b). Let τ (tau) be such a limit point. We say that

$$\varphi(\tau) + i\psi(\tau) = \xi + i\eta,$$

so that $P = \xi + i\eta$ lies on the curve. If $\varphi(\tau) + i\psi(\tau)$ were not equal to $\xi + i\eta$, then, as $\varphi(t)$ and $\psi(t)$ are continuous at τ, $\varphi(t_n) + i\psi(t_n)$ could not be close to $\xi + i\eta$ when t_n is close to τ. Thus, the points of the curve form a closed set.

We shall show now that if the curve has more than one point, every point of the curve is a limit point of points of the curve.

Let P_1 be any point of the curve. Let

$$P_1 = \varphi(t_1) + i\psi(t_1).$$

Let $P_2 = \varphi(t_2) + i\psi(t_2)$ be any point on the curve distinct from P_1. Suppose, to fix our ideas, that $t_2 > t_1$. Consider the function

$$F(t) = [\varphi(t) - \varphi(t_1)]^2 + [\psi(t) - \psi(t_1)]^2.$$

$F(t)$ is continuous for t on (a, b). We have

$$F(t_1) = 0, \quad F(t_2) = |P_2 - P_1|^2.$$

Hence, as t increases from t_1 to t_2, $F(t)$ takes on all values between 0 and $|P_2 - P_1|^2$, in particular all very small positive values. Now $F(t)$, for any t, equals $|Q - P_1|^2$, where Q is the point on the curve obtained from t. Thus there are points Q on the curve, distinct from P_1, with $|Q - P_1|$ as small as one pleases. Hence P_1 is a limit point of points on the curve.

Q.E.D.

EQUIVALENT CURVES

3. Consider a continuous curve

(1) $\qquad x = \varphi(t), \qquad y = \psi(t), \qquad a \leq t \leq b.$

Suppose that a continuous curve

(2) $\qquad x = \xi(\tau), \qquad y = \eta(\tau), \qquad c \leq \tau \leq d$

has the following relation to the curve (1):

There exists a function

$$\tau = f(t),$$

continuous and increasing on (a, b), with $f(a) = c$, $f(b) = d$ such that

$$\xi[f(t)] = \varphi(t), \qquad \eta[f(t)] = \psi(t),$$

for every t in (a, b).

We shall say that the curve (2) is *equivalent* to the curve (1).

If there exists a function

$$\tau = f(t)$$

continuous and decreasing on (a, b) with $f(a) = d$, $f(b) = c$, such that

$$\xi[f(t)] = \varphi(t), \qquad \eta[f(t)] = \psi(t),$$

we shall say that (2) is *inversely equivalent* to (1).

The result of §1 shows that if (2) is equivalent to (1), (1) is equivalent to (2). A similar result holds for inverse equivalence.

It is obvious that two curves which are equivalent or inversely equivalent have the same points.

SIMPLE CURVES

4. Consider a continuous curve

$$x = \varphi(t), \qquad y = \psi(t), \qquad a \leq t \leq b.$$

Suppose that no two distinct values of t give the same point on the curve; that is, that the equations

$$\varphi(t_2) = \varphi(t_1)$$
$$\psi(t_2) = \psi(t_1)$$

have no solution with $t_2 \neq t_1$. We shall call such a curve a *simple curve*.

REPRESENTATION OF A SIMPLE CURVE

5. Suppose that *two simple curves*

(1) $\qquad x = \varphi(t), \qquad y = \psi(t), \qquad a \leq t \leq b$

(2) $\qquad x = \xi(\tau), \qquad y = \eta(\tau), \qquad c \leq \tau \leq d,$

have the same points.

We shall show that the *two curves are either equivalent or inversely equivalent.*

Proof: Given any t of (a, b), there is precisely one τ of (c, d) which gives the same point which t does. This is because the curves (1) and (2) are simple.

Let the correspondence just described be given by

$$\tau = f(t).$$

If we can show that $f(t)$ is continuous, we shall know, because distinct t's give distinct τ's, that $f(t)$ is either increasing or decreasing. (See previous pages on monotonic functions.) This will prove our theorem.

Let C represent the points of the curves (1) and (2). For every $x + iy$ of C, there is a single τ. We write

$$\tau = \alpha(x, y).$$

We shall prove that τ is continuous in both variables on C, that is, for every $\varepsilon > 0$, we can find a $\delta > 0$ such that

$$|\alpha(x', y') - \alpha(x, y)| < \varepsilon$$

if $|x' - x| < \delta$ and $|y' - y| < \delta$. In this, $x + iy$ is supposed to be any fixed point of the curve.

Let us suppose that this is not so. Then, for some $\varepsilon > 0$, we can find an infinite sequence of points on the curve,

$$x_1 + iy_1, x_2 + iy_2, \ldots, x_n + iy_n, \ldots,$$

where x_n and y_n approach x and y respectively when n increases, such that

(3) $\qquad |\tau_n - \tau| > \varepsilon$

where $\tau_n = \alpha(x_n, y_n)$. The points τ_n have at least one limit point. Because of (3), such a limit point cannot be τ. Let $\tau' \neq \tau$ be a limit point. As

$$\varphi(\tau_n) = x_n, \quad \psi(\tau_n) = y_n,$$

we must have

$$\varphi(\tau') = x, \quad \psi(\tau') = y.$$

73

This contradicts the fact that (x, y) is found from only a single value of τ. Hence $\alpha(x, y)$ is continuous. Now

$$f(t) = \alpha[\varphi(t), \psi(t)].$$

Since, when t changes slightly, $\varphi(t)$ and $\psi(t)$ change slightly, a small change in t produces a small change in $\alpha[\varphi(t), \psi(t)]$. Thus, $f(t)$ is continuous.

Q.E.D.

XVI

Rectifiable Curves

SEGMENT AND LENGTH

1. Let (x_1, y_1) and (x_2, y_2) be any two points (that is, complex numbers) distinct or coincident. Of course (x_1, y_1) means $x_1 + iy_1$ and (x_2, y_2) means $x_2 + iy_2$. By the segment joining the two points, we mean the set of points.

$$[x_1 + t(x_2 - x_1), \quad y_1 + t(y_2 - y_1)]$$

where t takes on all values of the closed interval (0, 1).

By the *length* of the above segment, we mean

$$\sqrt{(x_2 - x_1)^2 + (y_2 - y_1)^2}.$$

INSCRIBED POLYGON

2. Consider a continuous curve

$$x = \varphi(t), \quad y = \psi(t) \quad a \leq t \leq b.$$

Let n be any positive integer.

Let n + 1 points

$$t_0 = a, \; t_1 > t_0, \; t_2 > t_1, \ldots, \; t_n = b$$

be taken on the interval (a, b). The points $t_1, t_2, \ldots, t_{n-1}$ may be taken arbitrarily, except that $t_i > t_{i-1}$, $i = 1, \ldots, n$.

Let $x_i = \varphi(t_i)$, $y_i = \psi(t_i)$, $i = 0, \ldots, n$.

For every i from 1 to n inclusive, we consider the segment joining (x_{i-1}, y_{i-1}) and (x_i, y_i). The points of all of these segments will be called a *polygon inscribed in the given curve*.

By the length of this polygon, we shall mean the sum of the lengths of the segments which form it, that is

$$\sum_{i=1}^{n} \sqrt{(x_i - x_{i-1})^2 + (y_i - y_{i-1})^2}$$

It is easy to prove that if new points of division are added to the interval (a, b), the new inscribed polygon will have a length not less than that of the first one. This follows from the fact that the modulus of the sum of two complex numbers does not exceed the sum of their moduli.

RECTIFIABLE CURVES

Let a curve be such that as the points of division t_i become more and more numerous, with the differences $t_i - t_{i-1}$ tending toward zero, the length of the inscribed polygon tends toward a limit. Explicitly, suppose that a number L (necessarily non-negative) exists such that, for every $\varepsilon > 0$ a $\delta > 0$ can be found such that, for every subdivision of (a, b) with $t_i - t_{i-1} < \delta$, $i = 1, \ldots, n$, we have for the length $\mathcal{L}$ of the inscribed polygon,

$$|L - £| < \varepsilon.$$

We shall call such a curve *rectifiable*.

We shall call L the *length* of the curve.

It is easily seen that if one of two equivalent or inversely equivalent curves is rectifiable, the other is also and the two curves have the same length.

CONDITION FOR RECTIFIABILITY

4. Consider the totality of polygons inscribed in a curve C. Each polygon has a length £.

Theorem: For C to be rectifiable, it is necessary and sufficient that the set of lengths £ for all possible polygons inscribed in C have an upper bound.

The condition means, of course, that a positive number G exists such that every length £ is less than G.

Proof: (a) *Necessity:* If we can find an inscribed polygon whose length is as large as we please, we can, by adding new points of division t_i to those which give this polygon, obtain an inscribed polygon of arbitrarily great length based on points of division t_i arbitrarily close together. This is because the addition of new points t_i cannot decrease the length. Thus, if the £'s are not bounded, the polygon lengths do not approach a limit as the points of division become numerous and close together.

(b) *Sufficiency:* Let L be the least upper bound of the lengths £. We shall show that C has a length and that the length of C is L. Let any $\varepsilon > 0$ be preassigned. Consider any particular subdivision $t_0, t_1, \ldots, t_n$ for which the corresponding polygon, call it P_0, has a length $£_0$ with $L - £_0 < \varepsilon$. Let $\delta > 0$ be such that, if $|t' - t| < \delta$, we have

$$[\varphi(t') - \varphi(t)]^2 + [\psi(t') - \psi(t)]^2 < \frac{\varepsilon^2}{n^2}$$

or, what is the same

$$\sqrt{[\varphi(t') - \varphi(t)]^2 + [\psi(t') - \psi(t)]^2} < \frac{\varepsilon}{n}.$$

Also, let δ be less than the least $t_i - t_{i-1}$ used for P_0.

Consider now any polygon P based on intervals (t_{i-1}, t_i) of length less than δ. Let £ be the length of P. Of course, $£ \leq L$.

Let the points of division which give P_0, and those which give P, be combined to give a single subdivision of (a, b). Let $\overline{P}$ be the polygon obtained from the finer subdivision and $\overline{£}$ its length. We have

$$\overline{£} \geq £, \quad \overline{£} \geq £_0.$$

We are going to estimate the size of $\overline{£} - £$. (Note the similarity of the present procedure to that employed in connection with the integrability of continuous functions.)

If $\overline{£} > £$, it is because certain of the points t_i of the subdivision for P_0 lie between two consecutive points of the subdivision for P. Because of the way in which δ was chosen there cannot be two points of the subdivision for P_0 between two consecutive points of the subdivision for P.

To fix our ideas, suppose that t_1 lies between the consecutive points τ_1 and τ_2 of the subdivision for P. The contribution caused by this to the excess of $\overline{£}$ over £ will be

$$\sqrt{[\varphi(t_1) - \varphi(\tau_1)]^2 + [\psi(t_1) - \psi(\tau_1)]^2} + \sqrt{[\varphi(\tau_2) - \varphi(t_1)]^2 + [\psi(\tau_2) - \psi(t_1)]^2}$$
$$- \sqrt{[\varphi(\tau_2) - \varphi(\tau_1)]^2 + [\psi(\tau_2) - \psi(\tau_1)]^2}$$

which does not exceed

(1) $\sqrt{[\varphi(t_1) - \varphi(\tau_1)]^2 + [\psi(t_1) - \psi(\tau_1)]^2} + \sqrt{[\varphi(\tau_2) - \varphi(t_1)]^2 + [\psi(\tau_2) - \psi(t_1)]^2}$.

As $\tau_2 - \tau_1 < \delta$, we have $t_1 - \tau_1 < \delta$, $\tau_2 - t_1 < \delta$.

Hence the expression (1) is less than

$$\frac{\varepsilon}{n} + \frac{\varepsilon}{n} = \frac{2\varepsilon}{n}$$

It is easy now to see, considering the points $t_1, \ldots, t_{n-1}$, that

$$\overline{\pounds} - \pounds < (n - 1)\frac{2\varepsilon}{n} < 2\varepsilon.$$

As $\overline{\pounds} \geq \pounds_0$, we have

$$\pounds_0 - \pounds < 2\varepsilon.$$

As $L - \pounds_0 < \varepsilon$, we have

$$L - \pounds < 3\varepsilon,$$

provided that the subdivision of (a, b) which gives P has all its intervals shorter than δ.

This proves that C is rectifiable and has L for length.

FUNCTIONS OF BOUNDED VARIATION

5. Consider a function

$$y = f(x) \qquad a \leq x \leq b.$$

Let n be any positive integer. Let points $x_0, \ldots, x_n$ be so chosen that

$$x_0 = a, \; x_1 > x_0, \; x_2 > x_1, \ldots, \; x_n = b.$$

Let $y_i = f(x_i)$, $i = 0, \ldots, n$. Consider the sums, for all possible n's and for all possible subdivisions,

$$|y_1 - y_0| + |y_2 - y_1| + \ldots + |y_n - y_{n-1}|.$$

If the totality of these sums is a bounded set of numbers, that is, if there exists a number G which exceeds every sum, we shall say that $f(x)$ is of *bounded variation* on (a, b).

Theorem: If $f(x)$ is monotonic on (a, b) then $f(x)$ is of bounded variation on (a, b).

Proof: To fix our ideas, let $f(x)$ be non-decreasing.

Then

$$y_i - y_{i-1} \geq 0, \; i = 1, \ldots, n$$

so that

$$|y_1 - y_0| + \ldots + |y_n - y_{n-1}| = y_1 - y_0 + y_2 - y_1 + \ldots + y_n - y_{n-1} = y_n - y_0 = f(b) - f(a),$$

so that $\Sigma |y_i - y_{i-1}|$ is certainly bounded.

Q.E.D.

Example:

$$y = \sqrt{a^2 - x^2} \qquad 0 \leq x \leq a$$

Theorem: If $f(x)$ has a derivative throughout (a, b) and if the derivative is bounded on (a, b), then $f(x)$ is of bounded variation on (a, b).

Proof: Let $f'(x)$ exist and let $|f'(x)| < G$ on (a, b). Then $y_i - y_{i-1} = f'(\bar{x})(x_i - x_{i-1})$. (Mean value theorem.) Here $\bar{x}$ is some point between x_{i-1} and x_i. Thus

$$|y_i - y_{i-1}| < G(x_i - x_{i-1}).$$

Hence

$$|y_1 - y_0| + \ldots + |y_n - y_{n-1}| < G[(x_1 - x_0) + \ldots + (x_n - x_{n-1})] = G(b - a).$$

Q.E.D.

RECTIFIABILITY AND FUNCTIONS OF BOUNDED VARIATION

Theorem: For the curve

$$x = \varphi(t), \qquad y = \psi(t), \qquad a \leq t \leq b$$

to be rectifiable, it is necessary and sufficient that $\varphi(t)$ and $\psi(t)$ be of bounded variation.

Proof: (a) *Necessity* - We have

$$\sqrt{(x_i - x_{i-1})^2 + (y_i - y_{i-1})^2} \geq |x_i - x_{i-1}|$$

so that

$$\sum_{i=1}^{n} \sqrt{(x_i - x_{i-1})^2 + (y_i - y_{i-1})^2} \geq \sum_{i=1}^{n} |x_i - x_{i-1}|,$$

from which it follows that if the lengths of the inscribed polygons have an upper bound, the sums

$$|x_1 - x_0| + \ldots + |x_n - x_{n-1}|$$

have an upper bound.

Thus, if the curve is rectifiable $x = \varphi(t)$ is of bounded variation. We treat $y = \psi(t)$ similarly.

Q.E.D.

(b) *Sufficiency* - We have, by elementary algebra,

$$\sqrt{(x_i - x_{i-1})^2 + (y_i - y_{i-1})^2} \leq |x_i - x_{i-1}| + |y_i - y_{i-1}|$$

which shows that if $\varphi(t)$ and $\psi(t)$ are of bounded variation, the curve is rectifiable.

XVII

Curvilinear Integrals

FUNCTIONS CONTINUOUS ON A CURVE

1. Consider a continuous curve

$$x = \varphi(t), \qquad y = \psi(t) \qquad a \leq t \leq b.$$

We may represent the curve, if we let $z = x + yi$, by the single equation

$$z = \varphi(t) + i\psi(t) \qquad a \leq t \leq b.$$

Let $F(t) = \alpha(t) + i\beta(t)$ be a complex function of the real variable t, defined on (a, b). We understand, of course, that $\alpha(t)$ and $\beta(t)$ are real functions of t.

For $F(t)$ to be continuous on (a, b), it is necessary and sufficient that $\alpha(t)$ and $\beta(t)$ be continuous on (a, b). (Continuity of $F(t)$ is defined in the expected manner.)

We are going to think of $F(t)$ as being a function of z, where z is a point of the curve. As a single point z of the curve may correspond to many different values of t, $F(t)$ need not necessarily lead to a single-valued function of z. Still, when we speak of a point of the curve, we shall understand the point to be identified with a definite single value of t. Thus, given any z of the curve, there will be a definite t furnished to us as being associated with the z, so that we can find $F(t)$. We shall use the symbol $f(z)$ for $F(t)$, and, when $F(t)$ is continuous on (a, b), we shall speak of $f(z)$ as being "continuous along the curve."

If the curve is simple, $f(z)$ is actually a one-valued function of z and the preceding remarks relative to the meaning of $f(z)$ become unnecessary.

CURVILINEAR INTEGRAL

2. Consider a continuous curve C, given by

$$z = \varphi(t) + i\psi(t) \qquad a \leq t \leq b$$

Let $f(z) = F(t)$ be a function defined along the curve. Let (a, b) be divided into n subintervals by points

$$t_0 = a, \ t_1 > t_0, \ldots, \ t_n = b.$$

Let $z_j = \varphi(t_j) + i\psi(t_j)$, $j = 0, \ldots, n$.

In each interval (t_{j-1}, t_j), we choose arbitrarily a point τ_j and let

$$\zeta_j = \varphi(\tau_j) + i\psi(\tau_j).$$

We form the sum

(1) $$f(\zeta_1)(z_1 - z_0) + \ldots + f(\zeta_n)(z_n - z_{n-1}).$$

By the *norm* of the sum (1), we mean the greatest of the quantities $t_j - t_{j-1}$, $j = 1, \ldots, n$.

If the sum (1) approaches a limit as its norm approaches zero, that is, if a complex number I exists such that, for every $\varepsilon > 0$, we can find a $\delta > 0$ such that, for any sum (1) of norm less than δ, we have

$$|I - \sum_{j=1}^{n} f(\zeta_j)(z_j - z_{j-1})| < \varepsilon.$$

we shall say that $f(z)$ is *integrable* along C. We shall call I the *integral along C* of $f(z)\,dz$. We write

$$I = \int_C f(z)\,dz.$$

CONDITION FOR INTEGRABILITY

3. *Theorem: For $f(z)$ to be integrable along C, it is necessary and sufficient that for every $\varepsilon > 0$ a $\delta > 0$ exist such that any two sums*

$$\sum_{j=1}^{n} f(\zeta_j)(z_j - z_{j-1})$$

of norms less than δ have a difference less than ε in modulus.

Proof: (a) *Necessity* - Clear.

(b) *Sufficiency* - Let

$$\varepsilon_1, \varepsilon_2, \ldots, \varepsilon_n, \ldots$$

be a sequence of positive numbers, converging to zero, with

$$\varepsilon_1 > \varepsilon_2 > \ldots > \varepsilon_n > \ldots$$

Let $\delta_1, \delta_2, \ldots, \delta_n, \ldots$ be a sequence of positive numbers such that

$$\delta_1 > \delta_2 > \ldots > \delta_n > \ldots$$

and such that any two sums of norms less than δ_n have a difference less than ε_n in modulus. Let

(2) $$\Sigma_1, \Sigma_2, \ldots, \Sigma_n, \ldots$$

be a sequence of sums for $f(z)$, Σ_n being of norm less than δ_n.

If $m > n$, $|\Sigma_m - \Sigma_n| < \varepsilon_n$. By Cauchy's convergence criterion for real sequences, the real parts of the numbers in (2) and also the coefficients of i, furnish convergent real sequences. By a theorem previously proved for sequences of complex numbers, (2) converges.

Let I be the limit of (2). We say that I has all the qualities necessary for it to be an integral of $f(z)\,dz$ along C.

Let any $\varepsilon > 0$ be assigned. Take n so that $\varepsilon_n < \varepsilon$. We keep n fixed. Take $\delta = \delta_n$. Let Σ be any sum of norm less than δ. We say that

$$|I - \Sigma| \leq \varepsilon.$$

If

$$|I - \Sigma| > \varepsilon,$$

we have, for m sufficiently large,

$$|\Sigma_m - \Sigma| > \varepsilon$$

where Σ_m is the mth term of (2). But, if $m > n$, both Σ_m and Σ will be of norm less than δ_n. (Note that $\delta_m < \delta_n$ and $\delta = \delta_n$.) Hence, we will have

$$|\Sigma_m - \Sigma| < \varepsilon_n < \varepsilon.$$

This contradiction proves the theorem.

EXISTENCE THEOREM

4. Theorem: Let C, given by

$$z = \varphi(t) + i\psi(t), \qquad a \leq t \leq b,$$

be a rectifiable curve. Let $f(z)$ be continuous along C. Then $f(z)$ is integrable along C.

Proof: We have $f(z) = F(t)$, where $F(t)$, being continuous on the closed interval (a, b), is uniformly continuous on that interval.

Let the length of C be L.

Given an $\varepsilon > 0$, let $\delta > 0$ be such that

$$|F(t_2) - F(t_1)| < \frac{\varepsilon}{2L}$$

if $|t_2 - t_1| < \delta$. We say that any two sums of norms less than δ have a difference which is less than ε in modulus.

Let Σ_1 and Σ_2 be two such sums. We form a sum Σ based on intervals obtained by combining the points of division used in Σ_1 and those used in Σ_2.

We shall estimate $|\Sigma - \Sigma_1|$. Let (t_{i-1}, t_i) be any interval used in Σ_1. Suppose that, when the points of division for Σ_2 are added, this interval has the points of division $s_0, s_1, \ldots, s_m$. Let τ_i be the value of t chosen in (t_{i-1}, t_i) in forming Σ_1 and let $\sigma_1, \ldots, \sigma_m$ be the values of t chosen in $(s_0, s_1), \ldots, (s_{m-1}, s_m)$ in forming Σ. Then the terms of Σ which come from (t_{i-1}, t_i) have the sum

$$(3) \qquad F(\sigma_1)(z_{s_1} - z_{s_0}) + \ldots + F(\sigma_m)(z_{s_m} - z_{s_{m-1}})$$

where

$$z_{s_j} = \varphi(s_j) + i\psi(s_j), \; j = 0, \ldots, m.$$

The difference between (3) and $F(\tau_i)(z_{s_m} - z_{s_0})$ may be written

$$(4) \quad [F(\sigma_1) - F(\tau_i)](z_{s_1} - z_{s_0}) + \ldots + [F(\sigma_m) - F(\tau_i)](z_{s_m} - z_{s_{m-1}}).$$

As

$$|\sigma_j - \tau_i| < \delta, \; j = 1, \ldots, m,$$

we have

$$|F(\sigma_j) - F(\tau_i)| < \frac{\varepsilon}{2L}, \; j = 1, \ldots, m,$$

so that the quantity (4) has a modulus less than

$$(5) \qquad \frac{\varepsilon}{2L} [|z_{s_1} - z_{s_0}| + \ldots + |z_{s_m} - z_{s_{m-1}}|].$$

Now the quantity in brackets in (5) is the length of a polygon inscribed in that part of C which corresponds to $t_{i-1} \leq t \leq t_i$.

Hence, adding up expressions similar to (5), we find that

$$|\Sigma - \Sigma_1| < \frac{\varepsilon}{2L} \text{ [length of a polygon inscribed in C]}.$$

Then

$$|\Sigma - \Sigma_1| < \frac{\epsilon}{2L} \cdot L = \frac{\epsilon}{2}.$$

Similarly,
$$|\Sigma - \Sigma_2| < \frac{\epsilon}{2},$$

so that
$$|\Sigma_1 - \Sigma_2| < \epsilon.$$

Q.E.D.

AN EXAMPLE

5. Let C be a rectifiable curve which joins two points A and B. We say that
$$\int_C z\, dz = \frac{B^2 - A^2}{2}$$

What precedes implies that z is integrable along any rectifiable curve.

Let
$$z_0 = A,\ z_1,\ z_2, \ldots,\ z_n = B$$

be points on the curve which correspond to a set of increasing values of the parameter t. We consider the two sums

$$I_1 = z_0(z_1 - z_0) + z_1(z_2 - z_1) + \ldots + z_{n-1}(z_n - z_{n-1}),$$
$$I_2 = z_1(z_1 - z_0) + z_2(z_2 - z_1) + \ldots + z_n(z_n - z_{n-1}).$$

We have, identically,
$$I_1 + I_2 = z_n^2 - z_0^2 = B^2 - A^2.$$

As n increases and the points z_i come closer together, I_1 and I_2 approach $\int_C z\, dz$. Hence
$$\int_C z\, dz = \frac{B^2 - A^2}{2}.$$

XVIII

Jordan Curves

DEFINITION

1. Consider a continuous curve
$$x = \varphi(t), \qquad y = \psi(t), \qquad a \leq t \leq b,$$
which has the following properties:

(A) $$\varphi(b) = \varphi(a); \qquad \psi(b) = \psi(a).$$

(B) The only solutions of the equations:
$$\varphi(t_2) = \varphi(t_1); \qquad \psi(t_2) = \psi(t_1)$$
with $t_1 \neq t_2$ are $t_1 = a, t_2 = b$ and $t_1 = b, t_2 = a$.

Briefly, two distinct points of (a, b), of which one at least is not an end point of (a, b), give two distinct points of the curve. The end-points of (a, b) give the same point of the curve.

We shall call such a curve a *simple closed curve* or a *Jordan curve*.

EQUIVALENT JORDAN CURVES

2. We are going to investigate the circumstances under which two Jordan curves

(1) $\qquad x = \varphi(t), \qquad y = \psi(t), \qquad a \leq t \leq b$

(2) $\qquad x = \xi(\tau), \qquad y = \eta(\tau), \qquad c \leq \tau \leq d$

can have the same points.

Case 1: Suppose that (1) and (2) have the same initial-terminal point, that is, that
$$\varphi(a) = \varphi(b) = \xi(c) = \xi(d)$$
$$\psi(a) = \psi(b) = \eta(c) = \eta(d).$$

We shall prove, in this case, that *(1) and (2) are either equivalent or inversely equivalent*.

Consider any t distinct from a and b. There corresponds to it a τ which gives the same point of (2) as the t does of (1). This defines a function
$$\tau = f(t)$$
for $a < t < b$.

We shall show first that $f(t)$ is continuous for $a < t < b$.

Consider any value t_0 of t. Suppose that for some $\varepsilon > 0$ it is impossible to find a $\delta > 0$ such that
$$|f(t) - f(t_0)| < \varepsilon$$
for $|t - t_0| < \delta$. Then there exists a sequence of t's,
$$t_1, t_2, \ldots, t_n, \ldots,$$

which approach t_0 such that, letting

$$\tau_0 = f(t_0); \quad \tau_n = f(t_n), \quad n = 1, 2, \ldots,$$

we have

$$|\tau_n - \tau_0| \geq \varepsilon.$$

The numbers

(3) $$\tau_1, \tau_2, \ldots, \tau_n, \ldots$$

are distinct from one another. This is because distinct t's in the interior of (a, b) give distinct points of the curve (1).

Hence the τ's of (3) have at least one limit point and τ_0 cannot be such a limit point.

Let τ' be a limit point of the τ's. Let

(4) $$\tau_{i_1}, \tau_{i_2}, \ldots$$

be a sequence of τ's approaching τ'. The sequence

(5) $$t_{i_1}, t_{i_2}, \ldots$$

converges to t_0.

The points of the curve (1) which correspond to the t's of sequence (5) thus tend toward the point

$$[\varphi(t_0), \psi(t_0)]$$

as a limit. The points of the curve (2) which correspond to the τ's of (4) approach this same point as a limit. But the points of curve (2) which are obtained from (4) approach

$$[\xi(\tau'), \eta(\tau')].$$

Hence the point $[\xi(\tau_0), \eta(\tau_0)]$ must be identical with $[\xi(\tau'), \eta(\tau')]$. This contradicts the fact that (2) is a simple closed curve and that τ_0 is neither c nor d.

Thus $f(t)$ is continuous for any t such that $a < t < b$.

We shall study the behavior of $f(t)$ as t decreases toward a. We shall prove first that $f(t)$ tends toward a limit, which is either c or d. Similarly, one will see that as t approaches b, $f(t)$ approaches one of the quantities c or d as a limit.

In any closed interval interior to (a, b), $f(t)$ is either an increasing function or a decreasing function, because it is continuous and assumes no value twice. It is easy to see, on this basis, that $f(t)$ is either increasing or decreasing throughout the open interval (a, b).

To fix our ideas, let us assume that $f(t)$ is increasing. We shall prove that as t approaches a, $f(t)$ approaches c.

The values which $f(t)$ assumes on the open interval (a, b) are the numbers of the open interval (c, d). This is because every τ of the open interval (c, d) correspond to some point of curve (2) which is distinct from the initial-terminal point.

Hence the lower bound of $f(t)$ on the open interval (a, b) is c.

Suppose that an $\varepsilon > 0$ exists for which no $\delta > 0$ can be found such that $|f(t) - c| < \varepsilon$ for $t - a < \delta$. Then, for a sequence of t's

$$t_1, t_2, \ldots, t_n, \ldots$$

approaching a, we have $f(t_n) \geq c + \varepsilon$. Now, given any t in the open interval (a, b), there is some t_n above less than t. As $f(t)$ is increasing, we have $f(t) > c + \varepsilon$ for every t, which contradicts the fact that that c is the lower bound of $f(t)$ for the open interval (a, b).

Hence, as t approaches a, f(t) approaches c. Similarly, as t approaches b, f(t) approaches d.

We now define f(t) as equal to c at a and equal to d at b. This makes f(t) continuous and increasing in (a, b).

For every t of the closed interval (a, b), we have

$$\xi[f(t)] = \varphi(t), \quad \eta[f(t)] = \psi(t).$$

Hence (1) and (2) are equivalent.

Similarly, when f(t) is decreasing in the open interval (a, b), (1) and (2) are inversely equivalent.

Case 2. The initial-terminal points of (1) and (2) are not the same.

Suppose that e, distinct from c and d, is the value of τ which gives the initial-terminal point of (1). That is,

$$\xi(e) = \varphi(a) = \varphi(b)$$
$$\eta(e) = \psi(a) = \psi(b).$$

We consider a third curve, which we define as follows:

Let

(6) $\quad\quad x = \alpha(\tau), \quad y = \beta(\tau), \quad e \leq \tau \leq d + e - c,$

where, for $e \leq \tau \leq d$,

$$\alpha(\tau) = \xi(\tau), \quad \beta(\tau) = \eta(\tau)$$

and, for $d \leq \tau \leq d + e - c$,

$$\alpha(\tau) = \xi(\tau - d + c), \quad \beta(\tau) = \eta(\tau - d + c).$$

It is clear that (6) is a Jordan curve which has the same points as (2) and which has the same initial-terminal point as (1). Then (6) is either equivalent or inversely equivalent to (1).

We shall loosen the word *equivalent*, calling (2) equivalent or inversely equivalent to (1) according as (6) is equivalent or inversely equivalent to (1).

We can now say that two Jordan curves which have the same points are either equivalent or inversely equivalent.

APPLICATION TO INTEGRATION

3. Let f(z) be a function analytic in an open region. Consider any rectifiable Jordan curve C, whose points all lie in the open region. Then f(z) is integrable along C.

It follows easily from the preceding discussion that if another Jordan curve C' has the same points as C, then the integral of f(z) along C' is either equal to, or is the negative of, the integral along C. We have

$$\int_C = \int_{C'}$$

if C and C' are equivalent and

$$\int_C = -\int_{C'}$$

if C and C' are inversely equivalent.

XIX

Analysis Situs of the Triangle

STATEMENT OF THE JORDAN SEPARATION THEOREM

1. Let
$$z = \varphi(t) + i\psi(t), \qquad a \leq t \leq b$$
be any Jordan curve (simple closed curve).

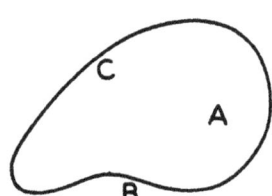

Let C represent the points of this curve. Let U represent the totality of complex numbers not belonging to C. The *Jordan Separation Theorem* states that U is made up of two open regions, A and B, which are such that any continuous curve which joins a point of A to a point of B contains at least one point of C. Of the two open regions A and B, one is bounded and the other is not. That one of the two open regions which is bounded is called the *interior* of C. The unbounded open region is called the *exterior* of C.

For proofs of the Jordan separation theorem as just stated, see Dienes, "The Taylor Series" or Kerekjarto, "Topologie." Restricted cases are treated in Osgood and in Tannery.

In what follows, we shall prove the separation theorem for the case in which C is a "triangle."

STRAIGHT LINES

2. Consider any equation $Ax + By + C = 0$ in which A, B and C are given *real* numbers with A and B not both zero. The totality of complex numbers $z = x + iy$ for which x and y satisfy the equation will be called a *straight line*.

Given any two distinct complex numbers z_1 and z_2, there is precisely one straight line which contains both of them. Two straight lines which are not identical have at most one point in common.

Let
$$z_1 = x_1 + y_1 i, \qquad z_2 = x_2 + y_2 i, \qquad z_3 = x_3 + y_3 i$$
be three complex numbers. For a straight line to exist which contains all three of these numbers, it is necessary and sufficient that
$$\begin{vmatrix} 1 & 1 & 1 \\ x_1 & x_2 & x_3 \\ y_1 & y_2 & y_3 \end{vmatrix} = 0$$

We have already defined the *segment* $\overline{z_1 z_2}$ joining two distinct points z_1 and z_2. It is the continuous curve
$$z = (1-t)z_1 + t z_2 \qquad 0 \leq t \leq 1.$$

It is easy to prove that every point of this segment lies on the line which contains z_1 and z_2.

TRIANGLES

3. Let z_1, z_2, z_3 be any three points which are not on a single straight line. The set of points consisting of the points of the three segments z_1z_2, z_2z_3, z_3z_1 will be called a *triangle*. A triangle consists of the points of a class of Jordan curves any two of which are equivalent or inversely equivalent. For instance, one such Jordan curve is

$$z = z_1 + t(z_2 - z_1) \qquad 0 \leq t \leq 1$$
$$z = z_2 + (t - 1)(z_3 - z_2) \qquad 1 \leq t \leq 2$$
$$z = z_3 + (t - 2)(z_1 - z_3) \qquad 2 \leq t \leq 3.$$

INTERIOR AND EXTERIOR

4. Consider any triangle T. Let U be the totality of points not belonging to T. We shall show that U is made up of two open regions, A and B, which are such that any continuous curve which joins a point of A to a point of B contains at least one point of T.

Of the two open regions A and B, it will turn out that one is bounded and the other is not. We shall call the bounded open region the *interior* of T and the unbounded one the *exterior* of T.

Proof: We show first that every point $z = x + yi$ of the complex plane has a unique representation of the type

$$z = t_1 z_1 + t_2 z_2 + t_3 z_3$$

where t_1, t_2, t_3 are real numbers such that

$$t_1 + t_2 + t_3 = 1.$$

This follows from the fact that if we represent z_j by $x_j + iy_j$, $j = 1, 2, 3$, the equations

$$1 = t_1 + t_2 + t_3$$
$$x = t_1 x_1 + t_2 x_2 + t_3 x_3$$
$$y = t_1 y_1 + t_2 y_2 + t_3 y_3$$

whose determinant

$$\begin{vmatrix} 1 & 1 & 1 \\ x_1 & x_2 & x_3 \\ y_1 & y_2 & y_3 \end{vmatrix}$$

is known not to be zero, has one and only one solution.

It is easy to see that T consists of those points z for which no t is negative and for which at least one t is zero. For instance, let $t_3 = 0$, $t_1 \geq 0$, $t_2 \geq 0$. Then

$$z = t_1 z_1 + t_2 z_2, \qquad t_1 + t_2 = 1$$

so that

$$z = (1 - t_2) z_1 + t_2 z_2$$

where $t_2 \geq 0$ and where, since $t_1 \geq 0$, we have $t_2 \leq 1$. Hence z is on the segment $\overline{z_1 z_2}$. Conversely, a point on the triangle has at least one t equal to zero and no t negative.

Thus, U consists of those points for which every t is positive and those points for which at least one t is negative.

Let A be the set of points for which every t is positive and let B be the set of points for which some t is negative.

We shall prove that A and B are open regions and that A is bounded.

The equations
$$t_1 + t_2 + t_3 = 1$$
$$x_1 t_1 + x_2 t_2 + x_3 t_3 = x$$
$$y_1 t_1 + y_2 t_2 + y_3 t_3 = y$$

define t_1, t_2, t_3 as linear functions of x and y; that is, the equations give
$$t_j = A_j + B_j x + C_j y, \qquad j = 1, 2, 3,$$

where the A_j, B_j, C_j are real constants. Suppose that, for some point $z_0 = x_0 + iy_0$, t_1, t_2, t_3 are all positive. Since the t's are continuous functions of x and y, any $z = x + iy$ close to z_0 will give positive t's. This shows that *every point of A is an interior point of A*. Similarly, if, for some z_0, some t is negative, that t will be negative for any z close to z_0. Hence *every point of B is an interior point of B*.

We shall now prove that if u and v are two points of A, the segment $\overline{uv}$ belongs to A. This will prove *that A is an open region*.

Let
$$u = r_1 z_1 + r_2 z_2 + r_3 z_3 \quad (r_1 + r_2 + r_3 = 1)$$
$$v = s_1 z_1 + s_2 z_2 + s_3 z_3 \quad (s_1 + s_2 + s_3 = 1)$$

Then, if $z = (1 - t) u + tv$ with $0 \leq t \leq 1$, we have

(1) $\quad z = [(1 - t) r_1 + ts_1] z_1 + [(1 - t) r_2 + ts_2] z_2 + [(1 - t) r_3 + ts_3] z_3.$

As t and $(1 - t)$ are non-negative, with at least one of them positive, and as every r_j and s_j is positive, the coefficients of z_1, z_2, z_3 in (1) are all positive. The sum of those coefficients is unity. Hence every z on $\overline{uv}$ lies in A. Thus, A is an open region.

We shall show that A is bounded. If
$$z = t_1 z_1 + t_2 z_2 + t_3 z_3$$
with $t_1 > 0$, $t_2 > 0$, $t_3 > 0$ and $t_1 + t_2 + t_3 = 1$, we have $t_1 < 1$, $t_2 < 1$, $t_3 < 1$. Then
$$|z| \leq t_1 |z_1| + t_2 |z_2| + t_3 |z_3| \leq |z_1| + |z_2| + |z_3|.$$

This proves the boundedness of A.

We prove now that B is an open region.

Suppose that u and v are two points belonging to B. Let
$$u = r_1 z_1 + r_2 z_2 + r_3 z_3 \qquad (r_1 + r_2 + r_3 = 1)$$
$$v = s_1 z_1 + s_2 z_2 + s_3 z_3 \qquad (s_1 + s_2 + s_3 = 1).$$

To fix our ideas, suppose that $r_1 < 0$, $s_2 < 0$. The curve

(2) $\quad z = r_1 z_1 + [r_2(1 - t) + s_2 t] z_2 + [r_3 + r_2 t - s_2 t] z_3 \qquad 0 \leq t \leq 1$

joins u to the point
$$w = r_1 z_1 + s_2 z_2 + (r_3 + r_2 - s_2) z_3.$$

Furthermore, as $r_1 < 0$, every point of (2) lies in B. (Note that the sum of the coefficients of z_1, z_2, z_3 in (2) is unity.)

Again the curve
$$z = [r_1(1 - t) + s_1 t] z_1 + s_2 z_2 + (r_3 + r_2 - s_2 + r_1 t - s_1 t) z_3 \qquad 0 \leq t \leq 1,$$
which lies in B, because $s_2 < 0$, joins w to
$$s_1 z_1 + s_2 z_2 + (r_3 + r_2 + r_1 - s_2 - s_1) z_3 = s_1 z_1 + s_2 z_2 + s_3 z_3 = v.$$

Hence u can be joined to v by a continuous curve lying in B. All other cases can be treated as above. Thus B is an open region.

Suppose now that u lies in A and v in B.

Representing u and v as above, let us, to fix our ideas, suppose $s_1 < 0$.

Consider any continuous curve

$$z = \varphi(\tau) + i\psi(\tau) \qquad a \leq \tau \leq b,$$

which joins u to v. Every z of this curve has associated with it a unique set t_1, t_2, t_3 which, as was seen above, are continuous functions of z. Hence t_1, t_2, t_3 are continuous functions of τ on (a, b). Explicitly, t_j is of the form

$$A_j + B_j \varphi(\tau) + C_j \psi(\tau), \qquad j = 1, 2, 3.$$

For τ slightly greater than a, t_1, t_2, t_3 will all be plus. For $\tau = b$, t_1 is minus.

There exist thus numbers ζ with $a < \zeta < b$ such that, for every τ with $a < \tau < \zeta$, t_1, t_2, t_3 are all positive. Let τ_0 be the least upper bound of the numbers ζ. Then $\tau_0 > a$. Because t_1 is negative for τ slightly less than b, we have $\tau_0 < b$.

No t_j can be negative for τ_0. Otherwise some t_j would be negative for τ slightly less than τ_0. Not every t_j can be positive for τ_0. Otherwise every t_j would be positive for τ slightly greater than τ_0 and τ_0 would not be an upper bound for the numbers ζ.

Hence at τ_0, at least one t_j is zero and the others non-negative. That is, $\varphi(\tau_0) + i\psi(\tau_0)$ is on T.

Since A and T are both bounded, B cannot be bounded.

This proves our theorem completely. We observe that it can easily be shown that T is the totality of boundary points either for A or for B.

DECOMPOSITION INTO FOUR TRIANGLES

5. Consider a triangle of vertices z_1, z_2, z_3. Let α be a point of $\overline{z_2 z_3}$ distinct from z_2 and z_3, β a point of $\overline{z_1 z_3}$ distinct from z_1 and z_3, γ a point of $\overline{z_1 z_2}$ distinct from z_1 and z_2.

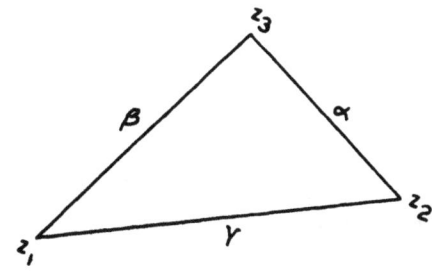

It is easy to prove that z_1, β and γ, for instance, are not collinear. Consider then the four triangles whose vertices are respectively

$$z_1 \ \beta \ \gamma$$
$$z_2 \ \alpha \ \gamma$$
$$z_3 \ \alpha \ \beta$$
$$\alpha \ \beta \ \gamma$$

We say that any point *on* any of these triangles is either *on* or *within* the triangle $z_1 z_2 z_3$ and that every point *within* any of these triangles is *within* the triangle $z_1 z_2 z_3$.

Consider, for instance, the triangle $z_1 \beta \gamma$. Let

$$\beta = (1 - r) z_1 + r z_3, \qquad 0 < r < 1$$
$$\gamma = (1 - s) z_1 + s z_2, \qquad 0 < s < 1.$$

Any point on or within $z_1 \beta \gamma$ is given by

$$z = t_1 z_1 + t_2 [(1 - r) z_1 + r z_3] + t_3 [(1 - s) z_1 + s z_2]$$

$$t_1 + t_2 + t_3 = 1, \ t_1 \geq 0, \ t_2 \geq 0, \ t_3 \geq 0.$$

For such a z, we have

$$z = z_1 [t_1 + t_2 (1 - r) + t_3 (1 - s)] + s t_3 z_2 + r t_2 z_3$$

or

(3) $$z = z_1 (1 - r t_2 - s t_3) + s t_3 z_2 + r t_2 z_3.$$

The sum of the coefficients of z_1, z_2, z_3 in (3) is unity. Neither $s t_3$ nor $r t_2$ is negative. Furthermore, as $r < 1$ and $s < 1$, we have

$$1 - r t_2 - s t_3 \geq 1 - t_2 - t_3 = t_1 \geq 0.$$

Thus, the coefficient of z_1 in (3) is non-negative.

The non-negative character of the coefficients in (3) proves the *on or within* case of the theorem.

Furthermore, if $t_1 > 0$, $t_2 > 0$, $t_3 > 0$ then $s t_3 > 0$ and $r t_2 > 0$, while $1 - r t_2 - s t_3 \geq t_1 > 0$, so that a point within $z_1 \beta \gamma$ is within $z_1 z_2 z_3$.

INTEGRATION

6. Let there be given an open region. Let a function $f(z)$ be *analytic* in this region, that is, possess a derivative at every point of the open region. Let z_1, z_2, z_3 be the vertices of a triangle every point on or within which lies in the given open region.

The symbol $z_1 z_2 z_3$ will now represent the Jordan curve

$$z = z_1 + t (z_2 - z_1) \qquad 0 \leq t \leq 1$$
$$z = z_2 + (t - 1) (z_3 - z_2) \qquad 1 \leq t \leq 2$$
$$z = z_3 + (t - 2) (z_1 - z_3) \qquad 2 \leq t \leq 3.$$

Of course, $f(z)$ is integrable along $z_1 z_2 z_3$

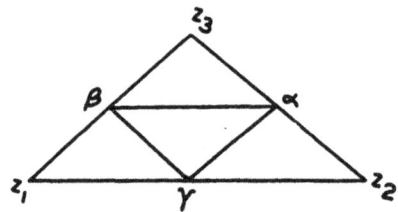

Let a subdivision of the given triangle into four triangles be made as in §5. It is not difficult to see, on the basis of earlier considerations relative to integration along inversely equivalent curves, that

$$\int_{z_1 z_2 z_3} f(z)\, dz = \int_{z_1 \gamma \beta} f(z)\, dz + \int_{\gamma z_2 \alpha} f(z)\, dz + \int_{\alpha z_3 \beta} f(z)\, dz + \int_{\alpha \beta \gamma} f(z)\, dz.$$

XX

The Cauchy Integral Theorem for Triangles

AN INEQUALITY FOR CURVILINEAR INTEGRALS

1. In what follows, we shall frequently use the term *curve* to mean *points of a curve*. No confusion will result.

Let C be a continuous curve. Let $f(z)$ be integrable along C. It is easy to prove, as in the cases of the real variable, that $f(z)$ is bounded on C. That is, an $M > 0$ exists such that

$$|f(z)| < M$$

for every z on C. This understood, we prove the important theorem:

Theorem: Let C be a rectifiable curve of length λ. Let $f(z)$ be a function integrable along C. Let $M \geq 0$ be such that $|f(z)| \leq M$ along C. Then

$$(1) \qquad \left|\int_C f(z)\, dz\right| \leq M\lambda.$$

Proof: Consider any sum of the type

$$\Sigma = f(\zeta_1)(z_1 - z_0) + \ldots + f(\zeta_n)(z_n - z_{n-1})$$

where the nature of the z_i and ζ_i is evident. We have, obviously,

$$|\Sigma| \leq M(|z_1 - z_0| + \ldots + |z_n - z_{n-1}|)$$

Now the quantity in parentheses is the length of a polygon inscribed in C. Hence

$$|\Sigma| \leq M\lambda.$$

This gives at once the inequality (1), for, if (1) did not hold, a Σ approximating with sufficient closeness to $\int_C f(z)\, dz$ would have a modulus greater than $M\lambda$.

STATEMENT OF THE CAUCHY INTEGRAL THEOREM FOR THE CASE OF A TRIANGLE

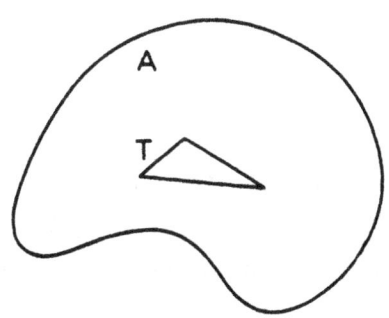

2. Let A be an open region. Let $f(z)$ be a function analytic throughout A. Let T be a triangle which lies, with its interior, in A. The theorem we are to prove states that

$$\int_T f(z)\, dz = 0.$$

We proceed to derive some results which will be useful in the proof of this theorem.

TWO CURVILINEAR INTEGRALS

3. Let B and C be two points of the complex plane, distinct or coincident.

We know from previous work that the integral of az, where a is any constant, along any rectifiable curve joining B and C is $\frac{a}{2}(C^2 - B^2)$. In particular, if B and C coincide, that is, if the curve is closed, the integral is zero.

91

With even simpler details, one can show that if a is any constant, the integral of a along any rectifiable curve joining B and C is a (C - B). In particular, if B and C coincide, the integral is zero.

ON DIFFERENTIABLE FUNCTIONS

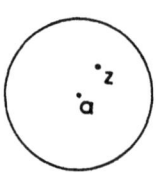

4. Let $f(z)$ be defined for a neighborhood of the point a and possess a derivative at a. We shall show that, in the given neighborhood of a, we have an expression for $f(z)$,

(2) $\qquad f(z) = f(a) + f'(a)(z - a) + \eta(z)(z - a)$

where $\eta(z)$ is a function of z which approaches 0 as z approaches a. That is, given any $\varepsilon > 0$, there is a $\delta > 0$ such that $|\eta(z)| < \varepsilon$ for $|z - a| < \delta$.

Proof: For $z = a$, we have (2) with $\eta(a) = 0$. Suppose that $z \neq a$. Let

(3) $\qquad \eta(z) = \dfrac{f(z) - f(a)}{z - a} - f'(a).$

Then $\eta(z)$ approaches 0 as z approaches a. Solving for $f(z)$ in (3), we find (2).

A THEOREM ON SEQUENCES OF TRIANGLES

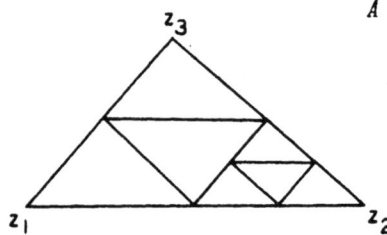

5. Let T be a triangle of vertices z_1, z_2, z_3. Using the midpoints of the sides of T, we find four triangles contained in T. Let T_1 be any one of these four triangles. Let T_1 be given the treatment accorded to T and let T_2 be any one of the four triangles contained in T_1. Continuing, we form an infinite sequence of triangles

$$T, T_1, T_2, \ldots, T_n, \ldots,$$

each containing the succeeding one.

We say that there exists one and only one point which has the property of lying on or within each triangle of the sequence.

Proof: Let λ be the perimeter of T, that is, let

$$\lambda = |z_2 - z_1| + |z_3 - z_2| + |z_1 - z_3|.$$

Then the perimeter of T_n is $\dfrac{\lambda}{2^n}$. Let z be any definite point on or within T, z_1 any point on or within T_1, and, in general, z_n any point on or within T_n. Consider the sequence

(4) $\qquad z, z_1, \ldots, z_n, \ldots.$

For any n and for any $p > 0$, z_{n+p} is on or within T_{n+p}, hence on or within T_n. Consequently $|z_{n+p} - z_n|$ is less than the perimeter of T_n. That is,

$$|z_{n+p} - z_n| < \dfrac{\lambda}{2^n}.$$

Our previous results on convergent sequences show us now that (4) converges to some limit P. It is evident that P cannot lie outside T. Otherwise (4) could not converge to P. Similarly, for every n, P is either on or within T_n. Furthermore, if Q is any point distinct from P, the perimeter of T_n will be less than $|Q - P|$ if n is large, so that, if n is large, Q cannot lie in T_n.

Q.E.D.

PROOF OF THE CAUCHY INTEGRAL THEOREM FOR TRIANGLES

Let $f(x)$ be analytic throughout the open region A. Let T be a triangle which lies, with its interior, in A. We are to prove that

$$\int_T f(z)\,dz = 0.$$

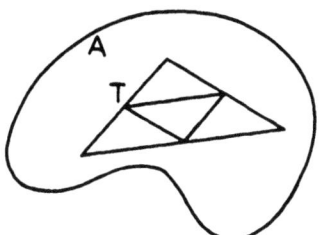

Proof: Let
$$M = \left|\int_T f(z)\,dz\right|.$$

What we have to show is that $M = 0$.

Let T be broken up into four triangles, in the usual way, by means of its midpoints. If those triangles are $\Delta_1, \ldots, \Delta_4$, we have
$$\int f(z)\,dz = \int_{\Delta_1} f(z)\,dz + \ldots + \int_{\Delta_4} f(z)\,dz.$$

Then

(5) $\qquad \left|\int_T\right| \leq \left|\int_{\Delta_1}\right| + \ldots + \left|\int_{\Delta_4}\right|.$

Hence, at least one of the quantities in the second member of (5) is not less than $\frac{M}{4}$.

Let then T_1 be one of the triangles $\Delta_1, \ldots \Delta_4$ such that
$$\left|\int_{T_1} f(z)\,dz\right| \geq \frac{M}{4}.$$

We treat T_1 as T was treated, finding a triangle T_2, contained in T_1, the perimeter of T_2 being half that of T_1 and T_2 being such that
$$\left|\int_{T_2} f(z)\,dz\right| \geq \frac{M}{4^2}$$

Continuing, we build an infinite sequence of triangles
$$T, T_1, T_2, \ldots, T_n, \ldots,$$

each containing its successor, the perimeter of T_n being $\lambda/2^n$ where λ is the perimeter of T, and the inequality

(6) $\qquad \left|\int_{T_n} f(z)\,dz\right| \geq \frac{M}{4^n}$

holding.

Let $\underline{a}$ be the point lying on or within every triangle of the sequence. We have, for z in A,

(7) $\qquad f(z) = f(a) + f'(a)(z-a) + \eta(z)(z-a)$

where $\eta(z)$ approaches 0 as z approaches a.

Let any $\varepsilon > 0$ be assigned. Let $\delta > 0$ be such that $|\eta(z)| < \varepsilon$ for $|z - a| < \delta$. Let n be so large that $\frac{\lambda}{2^n} < \delta$. Then, for z on T_n,
$$|z - a| < \lambda/2^n < \delta.$$

Thus, for z on T_n, we have $|\eta(z)| < \varepsilon$.

We observe that
$$\eta(z)(z-a) = f(z) - f(a) - f'(a)(z-a),$$

so that $\eta(z)(z-a)$ is continuous on T_n and hence integrable along T_n. By (7),
$$\int_{T_n} f(z)\,dz = \int_{T_n} f(a)\,dz + \int_{T_n} f'(a)(z-a)\,dz + \int_{T_n} \eta(z)(z-a)\,dz.$$

By §3, $\int_{T_n} f(a)\,dz = 0$ and
$$\int_{T_n} f'(a)(z-a)\,dz = \int_{T_n} f'(a)z\,dz - \int_{T_n} f'(a)a\,dz = 0.$$

Hence

(8) $$\int_{T_n} f(z) \, dz = \int_{T_n} \eta(z) \, (z - a) \, dz.$$

For z on T_n, we have $|\eta(z)| < \varepsilon$ and $|z - a| < \frac{\lambda}{2^n}$. Hence by §1,

(9) $$\left| \int_{T_n} \eta(z) \, (z - a) \, dz \right| \leq \varepsilon \cdot \frac{\lambda}{2^n} \cdot \frac{\lambda}{2^n} = \frac{\varepsilon \lambda^2}{4^n}$$

Comparing (6), (8) and (9), we see that

$$\frac{M}{4^n} \leq \frac{\varepsilon \lambda^2}{4^n},$$

so that $M \leq \varepsilon \lambda^2$. Since λ is fixed and ε can be taken arbitrarily small, it must be that $M = 0$.

Q.E.D.

XXI

Extension of the Cauchy Integral Theorem to Polygons

SIMPLY-CONNECTED OPEN REGIONS

1. An open region A is said to be *simply-connected* if, together with every Jordan curve C whose points lie in A, the interior of C lies in A. It is shown in the literature on analysis situs that the interior of a Jordan curve is simply connected.

POLYGONS

2. Let

(1) $$z_1, z_2, \ldots, z_n$$

be n points. We consider the continuous curve (made by taking in order the segments $\overline{z_i z_{i+1}}$),

(2)
$$z = z_1 + t(z_2 - z_1); \quad 0 \le t \le 1$$
$$z = z_2 + (t-1)(z_3 - z_2); \quad 1 \le t \le 2.$$
$$\ldots \ldots \ldots \ldots \ldots \ldots$$
$$z = z_{n-1} + (t - n + 2)(z_n - z_{n-1}); \quad n - 2 \le t \le n - 1.$$

The curve (2), or any curve equivalent to it, will be called *a polygon with the vertices* $z_1, \ldots, z_n$.

If $z_n = z_1$, the polygon (2) will be called *closed*. If, in addition, the polygon is a Jordan curve, it will be called a *simple closed polygon*.

STATEMENT OF THE CAUCHY INTEGRAL THEOREM FOR POLYGONS

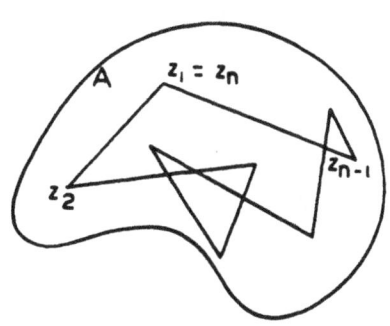

3. Let A be a simply connected open region. Let $f(z)$ be a function analytic throughout A.

Let C be any closed polygon lying in A. (That is, the points of C lie in A.)

We are to prove that

$$\int_C f(z) \, dz = 0.$$

For a closed polygon of two vertices (n = 3), the result is immediate because

$$\int_{\overline{z_1 z_2}} f(z) \, dz = - \int_{\overline{z_2 z_1}} f(z) \, dz.$$

Accordingly, in what follows, we limit ourselves to the case of $n \ge 4$, that is, to polygons of at least three sides.

REPLACEMENT OF THE POLYGON C BY A SIMPLER POLYGON

4. With respect to C, no generality is lost in assuming that *no consecutive pair of points z_i and z_{i+1} coincide*. When z_i and z_{i+1} coincide, we may suppress the segment $\overline{z_i z_{i+1}}$ as far as

integration is concerned. The assumption just indicated will apply to everything which follows.

This understood, let C be a polygon in A of n sides. In §5, where a proof by induction is made, it will be seen that we may limit ourselves to polygons C in which *no two of the points*

(1) $$z_1, z_2, \ldots, z_{n-1}$$

are coincident.

For a polygon C as just described, we shall prove that *there exists a polygon* C' *of* n *sides, lying in* A, *such that*

$$\int_C f(z)\, dz = \int_{C'} f(z)\, dz$$

and such that no pair of extremities of a side of C' *are collinear with any other vertex of* C'.

We begin by proving the lemma:

Lemma: Let A *be an open region. Let* $\overline{z_1 z_2}$ *be a segment lying in* A. *Then, if* z_3 *is sufficiently close to* z_2, *the segment* $\overline{z_1 z_3}$ *lies in* A.

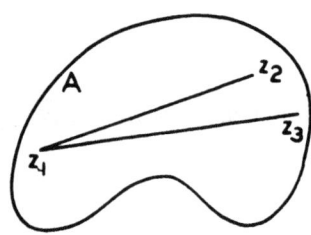

Proof: For every ζ on $\overline{z_1 z_2}$, there is a δ_ζ such that if $|z - \zeta| < \delta_\zeta$, z lies in A. About each ζ on $\overline{z_1 z_2}$ as center, let a circle of radius $\tfrac{1}{2}\delta_\zeta$ be described.

Since $\overline{z_1 z_2}$ is closed and bounded, there is a finite number of these circles such that every point on $\overline{z_1 z_2}$ is interior to at least one of them. Let δ be the least of the radii of the finite number of circles. Then, if ζ is on $\overline{z_1 z_2}$, and if $|z - \zeta| < \delta$, z is in A. For if ζ_1 is the center of one of the circles of the finite subset which contains ζ, then

(2) $$|\zeta - \zeta_1| < \tfrac{1}{2}\delta_{\zeta_1}$$

Also, if $|z - \zeta| < \delta$, we have, because $\delta \leq \tfrac{1}{2}\delta_{\zeta_1}$,

(3) $$|z - \zeta| < \tfrac{1}{2}\delta_{\zeta_1}.$$

By (2) and (3)

$$|z - \zeta_1| < \delta_{\zeta_1}$$

so that z lies in A.

We say now that if $|z_3 - z_2| < \delta$, then $\overline{z_1 z_3}$ lies in A. Every z on $\overline{z_1 z_3}$ is given by

$$z = z_1(1 - t) + tz_3, \qquad 0 \leq t \leq 1.$$

Thus

$$z = z_1(1 - t) + tz_2 + t(z_3 - z_2).$$

Let t be any given value. Then $z_1(1 - t) + tz_2$ is a definite point ζ on $\overline{z_1 z_2}$ and

$$z = \zeta + t(z_3 - z_2).$$

Thus
$$z - \zeta = t(z_3 - z_2),$$

and, as $|t| \leq 1$, and $|z_3 - z_2| < \delta$, we have

$$|z - \zeta| < \delta$$

so that z lies in A.

Q.E.D.

We return now to the replacement of the polygon C by the polygon C'.

Consider the segment $\overline{z_1 z_2}$. Suppose that certain vertices distinct from z_1 and z_2 are collinear with z_1 and z_2. It is easy to prove, on a rigorous, arithmetic basis, that in every neighborhood of z_2 there is a point $\underline{a}$ such that none of the vertices $z_3, z_4, \ldots, z_{n-1}$ is collinear with z_1 and $\underline{a}$. We can, furthermore, choose $\underline{a}$ so as not to be collinear with z_2 and z_3 or with z_1 and z_2.

If $\underline{a}$ is sufficiently close to z_2, the segments $\overline{z_1 a}$ and $\overline{a z_3}$ will lie in A. So also will $\overline{a z_2}$. Let C' be the polygon of vertices

$$z_1, a, z_3, \ldots, z_n = z_1.$$

We wish to show that

$$\int_{C'} f(z)\, dz = \int_C f(z)\, dz.$$

Evidently

(4) $$\int_{C'} - \int_C = \int_{\overline{z_1 a}} + \int_{\overline{a z_3}} - \int_{\overline{z_1 z_2}} - \int_{\overline{z_2 z_3}}$$

Now the triangle $z_1 a z_2$ lies in A. As A is simply connected, the interior of $z_1 a z_2$ lies in A. Hence

(5) $$\int_{\overline{a z_2}} + \int_{\overline{z_2 z_1}} + \int_{\overline{z_1 a}} = 0.$$

Similarly, considering the triangle $a z_2 z_3$, we have

(6) $$\int_{\overline{a z_2}} + \int_{\overline{z_2 z_3}} + \int_{\overline{z_3 a}} = 0.$$

Subtracting (6) from (5), we find

$$\int_{\overline{z_1 a}} - \int_{\overline{z_2 z_3}} + \int_{\overline{z_2 z_1}} - \int_{\overline{z_3 a}} = 0,$$

and as

$$\int_{\overline{a z_3}} = - \int_{\overline{z_3 a}}, \qquad \int_{\overline{z_1 z_2}} = - \int_{\overline{z_2 z_1}},$$

we see that the second member of (4) is zero and $\int_C = \int_{C'}$.

It is thus legitimate to assume that no vertex distinct from z_1 and z_2 is collinear with z_1 and z_2.

Suppose now that some vertex distinct from z_2 and z_3 is collinear with z_2 and z_3. Proceeding as above, we replace z_3 by a point close to z_3 so as to remove this condition. The point replacing z_3, being very close to z_3, will not be collinear with z_1 and z_2. Continuing in this manner, we secure a polygon C' lying in A, with no pair of extremities of a segment collinear with any other vertex. The integral of $f(z)$ along C' equals the integral along C.

COMPLETION OF PROOF

5. We now complete the proof of the theorem stated in §3.

If C has three sides, we may, as was seen in §4, replace C by a triangle. The Cauchy Integral Theorem is already proved for triangles, so that the case of three sides is disposed of.

Now, let the theorem be supposed proved for polygons of fewer than m sides, when m is some integer exceeding 3. We shall prove the theorem, by induction, for polygons of m sides.

First, let us consider the case, mentioned in §4, in which the points $z_1, z_2, \ldots, z_{n-1}$ are not all distinct. (Note that $n = m + 1$.)

Let, for instance, z_1 coincide with z_j, where j is some one of the integers $3, 4, \ldots, n-2$.

We consider the polygon C_1 whose vertices are

97

$$z_1, z_2, \ldots, z_j = z_1$$

and the polygon C_2 of vertices

$$z_j, z_{j+1}, \ldots, z_n = z_1 = z_j.$$

We have

$$\int_C f(z)\, dz = \int_{C_1} f(z)\, dz + \int_{C_2} f(z)\, dz.$$

As C_1 and C_2 have less than m sides, we have $\int_{C_1} = \int_{C_2} = 0$. Thus $\int_C f(z)\, dz = 0$.

We now consider the case in which the m vertices of C are distinct and replace C by a polygon C', of m sides, as in §4. We now distinguish two cases.

Case 1. C' is a simple closed polygon.

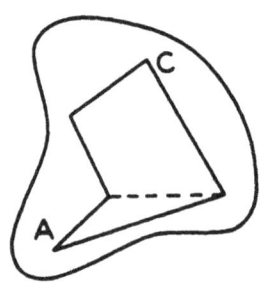

Let the vertices of C' be

$$z_1, z_2, z_3, \ldots, z_n = z_1.$$

A segment of the type $\overline{z_i z_j}$ where z_i and z_j are not the extremities of a side of C' will be called a diagonal of C'.

It is proved in the literature on analysis situs (see Kerekjarto, *Topologie*), that, given a polygon of more than three sides, with no two adjacent sides collinear, there is at least one diagonal of the polygon which lies, except for its extremities, within the polygon.

Plainly no two adjacent sides of C' are collinear so that C' has a diagonal which lies, except for its extremities, within C'. As any of the vertices of C' may be taken as the initial-terminal point of C', then some diagonal $\overline{z_1 z_j}$ where j is some integer among 3, 4, ..., n − 2, may be assumed to be of the type described above.

Consider the polygon C_1 whose vertices are

$$z_1, z_2, \ldots, z_j, z_1$$

and the polygon C_2 of vertices

$$z_1, z_j, z_{j+1}, \ldots, z_n = z_1.$$

Because every point on $\overline{z_1 z_j}$ is on or within C', and because A is simply connected, every point on $\overline{z_1 z_j}$ is in A. Thus C_1 and C_2 are in A and we have

$$\int_{C'} f(z)\, dz = \int_{C_1} f(z)\, dz + \int_{C_2} f(z)\, dz.$$

Now C_1 and C_2 have fewer than m sides. Thus,

$$\int_{C_1} = \int_{C_2} = 0,$$

so that $\int_{C'} = 0$ and $\int_C = 0$.

Case 2. C' is not simple.

Because C' satisfies the condition of §4, it must be that two non-adjacent sides of C' have a point in common. Such a point, by §4, cannot be an extremity of either of the sides. To fix our ideas, let us suppose that $\overline{z_1 z_2}$ has a point, called it ζ, in common with some $\overline{z_i z_{i+1}}$ with $2 < i < n - 1$. We consider the polygon C_1 of vertices

$$z_1, \zeta, z_{i+1}, z_{i+2}, \ldots, z_n = z_1,$$

and the polygon C_2 of vertices

$$\zeta, z_2, z_3, \ldots, z_i, \zeta.$$

We have

$$\int_{C'} = \int_{C_1} + \int_{C_2}.$$

As C_1 and C_2 have fewer than m sides, we have $\int_{C'} = 0$ and the proof is completed.

XXII

The Cauchy Integral Theorem for a Rectifiable Curve

CURVES IN OPEN REGIONS

1. Let A be an open region and C a continuous curve lying in A. We say that *there exists a $\delta > 0$ such that every z of C is the center of a circle of radius δ all points within which belong to A.*

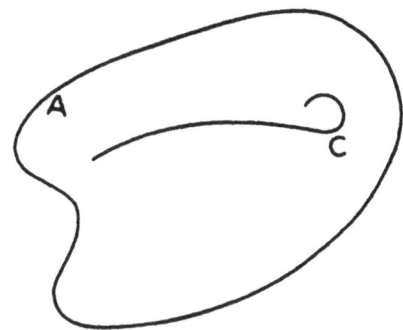

Proof: For every z of C, we can find a positive δ_z such that the circle interior

$$|z' - z| < \delta_z$$

contains only points z' belonging to A.

Let a circle of radius $\tfrac{1}{2}\delta_z$ be taken, with z as center, for every z of C. By Borel's theorem, there exist a finite number of these circle interiors which cover C. Taking such a finite set of circle interiors, let δ be the least of their radii.

Let z be any point of C. Let ζ be the center of some one of those circles of the finite set which contain z. Then $|z - \zeta| < \dfrac{\delta_\zeta}{2}$. Now, if $|z' - z| < \delta$, we have

$$|z' - \zeta| \leq |z' - z| + |z - \zeta| < \delta + \tfrac{1}{2}\delta_\zeta \leq \delta_\zeta.$$

Thus, z' is in A. That is, z is the center of a circle of radius δ, all points interior to which are in A.

OPEN REGIONS, CURVES AND APPROXIMATING POLYGONS

2. Let A be an open region and C a continuous curve lying in A. Let C have the representation

$$z = \varphi(t) + i\psi(t) \qquad a \leq t \leq b.$$

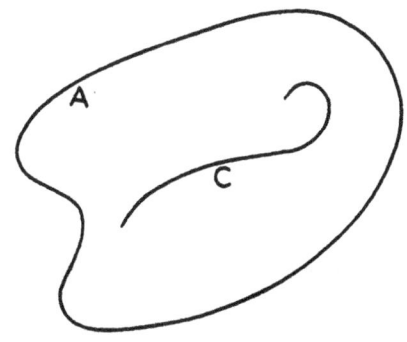

Let $t_0 = a$, $t_1 > t_0$, $t_2 > t_1$, ..., $t_n = b$, be $n + 1$ points on (a, b), which are arbitrary except for the conditions just placed upon them. Let

(1) $\qquad z_0, z_1, \ldots, z_n$

be the points $\varphi(t_j) + i\psi(t_j)$ on the curve C, $j = 0, \ldots, n$.

We consider the polygon (inscribed in C), whose vertices are the points (1). Let this polygon be represented by P. We say that *there exists an $\eta > 0$ such that, if $t_j - t_{j-1} < \eta$, $j = 1, \ldots, n$, every point on P lies in A.*

Proof: Let $\delta > 0$ be such that if $|z' - z| < \delta$, where z is on C, then z' is in A. The representation of C being, as above,

$$z = \varphi(t) + i\psi(t), \qquad a \leq t \leq b,$$

99

we see that z is a *uniformly* continuous function of t on (a, b). Let, then, $\eta > 0$ be such that if $|t' - t| < \eta$, then $|z' - z| < \delta$, where

$$z = \varphi(t) + i\psi(t), \quad z' = \varphi(t') + i\psi(t').$$

Consider any polygon P for which $|t_j - t_{j-1}| < \eta$, $j = 1, \ldots, n$. Let z be any point of P. Let $z_{j-1} z_j$ be one of the segments of P on which z lies. (Of course, z may lie on several segments.)

Then

$$|z - z_j| \leq |z_{j-1} - z_j| < \delta$$

so that z lies in A.

APPROXIMATION TO INTEGRALS

3. Let A be an open region. Let C be a *rectifiable* curve lying in A. Let f(z) be a function analytic throughout A. Then f(z) is integrable along C. Also f(z) is integrable along any polygon inscribed in C, if the polygon lies in A.

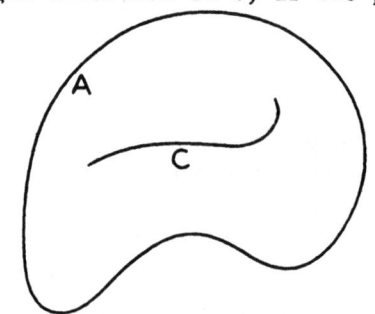

Let P be a polygon inscribed in C, based on points of division $t_0, t_1, \ldots, t_n$.

We say that, *given any $\varepsilon > 0$, we can find a $\delta > 0$ such that, if each $t_j - t_{j-1} < \delta$, then P lies in A and*

$$\left| \int_C f(z)\, dz - \int_P f(z)\, dz \right| < \varepsilon.$$

Proof: About each point of C as center, construct a circle which lies, interior and boundary, in A. Let a finite number of these circles be selected which cover C. The interior points of these circles form an open region. (Easy to prove.) Let U be this open region. The boundary points of U are points on the boundaries of the circles which gave rise to U. Let V be the set of points consisting of U and of the boundary points of U. Then V lies in A. Hence f(z) has a derivative at every point of V. A fortiori, f(z) is continuous at every point of V. Thus, as V is bounded, perfect set, f(z) is uniformly continuous throughout V.

Let L be the length of C.

Let any $\varepsilon > 0$ be assigned.

Let $\eta > 0$ be taken so that, if $|z' - z| < \eta$, with z' and z in V, then $|f(z') - f(z)| < \frac{\varepsilon}{2L}$.

Let $\delta > 0$ be chosen so that:

(1) If each $t_j - t_{j-1} < \delta$, then P lies in U.

(2) If $|t' - t| < \delta$, then $|z' - z| < \eta$, where

$$z = \varphi(t) + i\psi(t), \quad z' = \varphi(t') + i\psi(t').$$

We say that, *if each $t_j - t_{j-1} < \delta$, then*

$$\left| \int_C f(z)\, dz - \int_P f(z)\, dz \right| \leq \varepsilon.$$

Let $\widehat{z_{j-1} z_j}$ represent that part of C which corresponds to the interval (t_{j-1}, t_j), that is, the curve

$$z = \varphi(t) + i\psi(t), \quad t_{j-1} \leq t \leq t_j.$$

We have, for any z in A,

$$f(z) = f(z_{j-1}) + [f(z) - f(z_{j-1})],$$

so that, since $f(z_{j-1})$ is a constant,

(3) $\int_{\widehat{z_{j-1} z_j}} f(z)\, dz = f(z_{j-1})(z_j - z_{j-1}) + \int_{\widehat{z_{j-1} z_j}} [f(z) - f(z_{j-1})]\, dz.$

Similarly, in integrating along the segment $\overline{z_{j-1} z_j}$, we have

(4) $\int_{\overline{z_{j-1} z_j}} f(z)\, dz = f(z_{j-1})(z_j - z_{j-1}) + \int_{\overline{z_{j-1} z_j}} [f(z) - f(z_{j-1})]\, dz.$

Now, for any z on $\widehat{z_{j-1} z_j}$ or on $\overline{z_{j-1} z_j}$, we have $|z - z_{j-1}| < \eta$. Hence, for z on $\widehat{z_{j-1} z_j}$ or on $\overline{z_{j-1} z_j}$,

$$|f(z) - f(z_{j-1})| < \frac{\epsilon}{2L}.$$

Let λ be the length of $\widehat{z_{j-1} z_j}$. Then the length of $\overline{z_{j-1} z_j}$ does not exceed λ.

We have thus, by (3) and (4), after a subtraction,

$$\begin{aligned}\left|\int_{\widehat{z_{j-1} z_j}} f(z)\, dz - \int_{\overline{z_{j-1} z_j}} f(z)\, dz\right| &\leq \left|\int_{\widehat{z_{j-1} z_j}} [f(z) - f(z_{j-1})]\, dz\right| \\ &\quad + \left|\int_{\overline{z_{j-1} z_j}} [f(z) - f(z_{j-1})]\, dz\right| \\ &\leq \frac{\epsilon}{2L}\lambda + \frac{\epsilon}{2L}\lambda \\ &= \epsilon\frac{\lambda}{L}.\end{aligned}$$

From this it follows directly that

$$\left|\int_C f(z)\, dz - \int_P f(z)\, dz\right| \leq \epsilon.$$

THE CAUCHY INTEGRAL THEOREM FOR A RECTIFIABLE CURVE

4. *Theorem: Let A be a simply-connected open region. Let $f(z)$ be a function analytic throughout A. Let C be any closed rectifiable curve lying in A.*

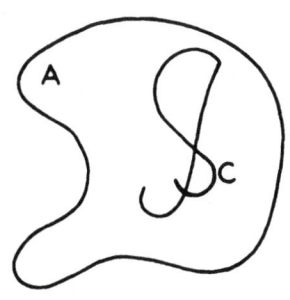

Then
$$\int_C f(z)\, dz = 0.$$

Remark: The case in which C is a Jordan curve has been especially prominent in the history of the theory of functions.

Proof: Let any $\epsilon > 0$ be taken. Let P be a closed polygon, inscribed in C and lying in A, such that

$$\left|\int_C f(z)\, dz - \int_P f(z)\, dz\right| < \epsilon.$$

As $\int_P f(z)\, dz = 0$, we have $\left|\int_C f(z)\, dz\right| < \epsilon$. Because ϵ is arbitrarily small, our theorem is proved.

Remarks on the Preceding Theorem:

5. As has already been observed, one of the most important cases is that in which C is a Jordan curve. Now, in this case, it is actually unnecessary to assume that A is simply-connected. A may be any open region provided that C is a rectifiable curve which lies, with its interior, in A.

Proofs of the theorem in this form, for restricted types of curves C, will be found in Osgood, *Funktionentheorie*, and in Tannery, *Théorie des fonctions*, Vol. II. The general case is treated in a paper by Kamke, *Mathematische Zeitschrift*, Vol. XXXV (1932), p. 539.

We shall, in certain later work, assume the truth of the theorem for the case just discussed.

XXIII

The Cauchy Integral Theorem for Several Contours

SENSED CIRCLES

1. Let $a + bi$ be any complex number. Let r be any positive real number. Consider the Jordan curve given by

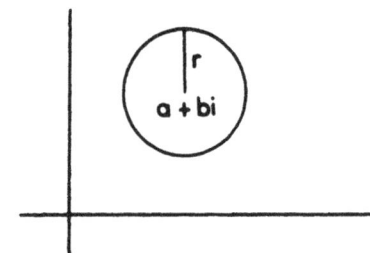

$$x = a + (r - t),\ y = b + \sqrt{r^2 - (r - t)^2},\ 0 \leq t \leq 2r$$
$$x = a + (t - 3r),\ y = b - \sqrt{r^2 - (t - 3r)^2},\ 2r \leq t \leq 4r.$$

It is easy to show that the points of this curve are those points z for which

$$|z - (a + bi)| = r.$$

Any curve *equivalent* to the above curve will be called a *positively sensed circle* with center at $(a + bi)$ and radius r. Any curve inversely equivalent to the above curve will be called a *negatively sensed circle* with center at $(a + bi)$ and radius r.

SENSE OF A JORDAN CURVE

2. Consider a Jordan curve C given by

(1) $\qquad\qquad x = \varphi(t), \qquad y = \psi(t), \qquad a \leq t \leq b.$

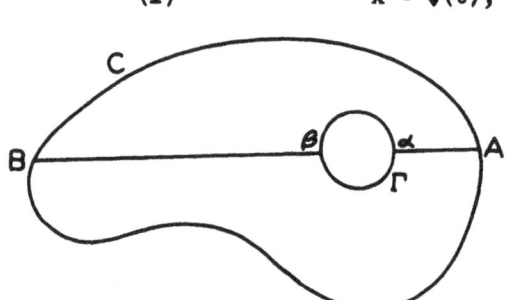

Let Γ be any *negatively* sensed circle which lies in the interior of C. It is proved in the theory of analysis situs that any point of Γ can be joined to any point of C by a continuous curve of which every point, with the exception of the extremities, is exterior to Γ and interior to C.

Let the rightmost point of Γ, call it α, be joined in this way to the initial point of C, which we shall represent by A. It is now possible to join the leftmost point of Γ, call it β, to any point B on C, distinct from A, by a simple curve of the above type which does not intersect the curve joining α to A.

Let the upper half of Γ, which is a continuous curve whose representation is obtained from the representation of Γ by using only the interval $(0, 2r)$, be denoted by Γ_1. Let the lower half of Γ be denoted by Γ_2.

The point B on C comes from a value t_0 of t in (1). Let C_1 be that part of C obtained when t varies from a to t_0 and let C_2 be that part obtained as t varies from t_0 to b.

Let $\widehat{\alpha A}$ be the simple curve mentioned above which joins α to A, its representation by functions being such that α is the initial point and A the terminal point. We denote by $\widehat{A\alpha}$ a curve inversely equivalent to $\widehat{\alpha A}$. We use similarly, with obvious significations, curves $\widehat{\beta B}$ and $\widehat{B\beta}$.

Now, let Δ_1 be the Jordan curve formed by uniting

$$\Gamma_1,\ \widehat{\alpha A},\ C_1,\ \widehat{B\beta}$$

and let Δ_2 be the Jordan curve formed by uniting

$$\Gamma_2,\ \widehat{\beta B},\ C_2,\ \widehat{A\alpha}.$$

It is a consequence of theorems of analysis situs that either the interior of Γ is *exterior* to Δ_1 and Δ_2 or else the interior of Γ is *interior* to Δ_1 and Δ_2.

In the first case, C is said to be *positively sensed* and, in the second case, C is said to be *negatively sensed*.

If a Jordan curve is positively sensed, every curve equivalent to it is positively sensed and every curve inversely equivalent to it is negatively sensed.

One might inquire as to the geometric (intuitive) attributes of sensed curves. If the Jordan curve C is positively sensed, one will find that as t increases from a to b, then $z = \varphi(t) + i\psi(t)$ moves in a counterclockwise direction, with the interior of C to the left.

INTEGRALS ALONG SENSED CURVES

3. In what follows immediately, we shall use the term *curve* to mean the *points of a curve*.

Let C be a rectifiable Jordan curve and let $f(z)$ be a function continuous on C. If we use a representation $z = \varphi(t) + i\psi(t)$ which gives C a positive sense and then integrate $f(z)$ along C, we shall say that we are *integrating $f(z)$ along C in the positive sense*. Similarly, we define *integration along C in the negative sense*.

REGIONS BOUNDED BY SEVERAL CONTOURS

4. Let C be a Jordan curve. Let $\Gamma_1, \Gamma_2, \ldots, \Gamma_n$ be n Jordan curves which are interior to C and exterior to each other. It is proved in analysis situs that the totality of points which are interior to C and exterior to $\Gamma_1, \ldots, \Gamma_n$, form an open region. The $n + 1$ curves will be the boundary of this open region.

Suppose now that C, $\Gamma_1, \ldots, \Gamma_n$ are rectifiable. Let $f(z)$ be a function defined on each of these curves and continuous along each of them. Let A be the open region with C, $\Gamma_1, \ldots, \Gamma_n$ for boundary.

By the *integral of $f(z)$ around the complete contour of* A, we mean the sum of the integral of $f(z)$ along C in the *positive* sense and of the integrals of $f(z)$ along the curves Γ_i in the *negative* sense.

THE CAUCHY INTEGRAL THEOREM FOR SEVERAL CONTOURS

5. Let A be an open region and $f(z)$ a function analytic throughout A. Let C, $\Gamma_1, \ldots, \Gamma_n$ be rectifiable curves lying in A, with $\Gamma_1, \ldots, \Gamma_n$ interior to C and exterior to each other. We suppose furthermore that every point of the open region B bounded by C, $\Gamma_1, \ldots, \Gamma_n$ lies in A.

The Cauchy Integral Theorem for several contours states that *the integral of $f(z)$ around the complete contour of B is zero*.

Proof: We shall limit ourselves to two contours, C and Γ, and shall use analysis situs on an intuitive basis. Let Γ be joined to C by two rectifiable curves, as in the accompanying figure. Then B is divided into two open regions, each bounded by a single rectifiable Jordan curve. Let C_1 and C_2 represent these two Jordan curves. Then the integral of $f(z)$ along the complete contour of B is easily seen to equal the sum of the integrals of $f(z)$ along C_1 and C_2, each of the latter integrals being taken in the positive sense.

The curve C_1 and its interior lie in A. By the strong form of the Cauchy Integral Theorem for a single contour, the integral of $f(z)$ along C_1 is zero. Similarly for C_2.

Q.E.D.

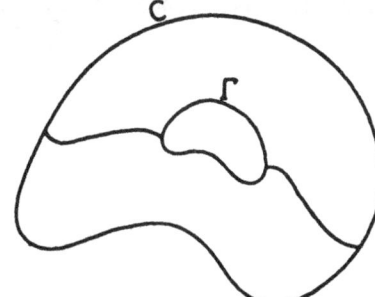

XXIV

Preliminaries for Cauchy Integral Formula

ANALYTICITY AT A POINT

1. A function $f(z)$ is said to be *analytic at the point* a if $f(z)$ is analytic in a *neighborhood* of the point $\underline{a}$, that is, if there is a circle with $\underline{a}$ as center throughout the interior of which $f(z)$ has a derivative.

COMBINATIONS OF ANALYTIC FUNCTIONS

2. Let $f(z)$ and $g(z)$ be analytic at $\underline{a}$. Then $f(z) + g(z)$ and $f(z) g(z)$ are analytic at $\underline{a}$. Also, if $g(a) \neq 0$, $\dfrac{f(z)}{g(z)}$ is analytic at $\underline{a}$. These results are obvious. In particular, if $\underline{a}$ is any constant $\dfrac{1}{z - a}$ is analytic at every point distinct from $\underline{a}$.

THE NUMBER π

3. A circle, positively or negatively sensed, is a rectifiable curve. We take as obvious the fact that the length of a circle is proportional to its radius. Thus, the ratio of the length of a circle to its radius is the same for all circles. We call half of this ratio π.

INTEGRAL OF $1/(z - a)$

4. Let C be a positively sensed circle, of any radius, with center at $\underline{a}$. We shall prove that

$$\int_C \frac{dz}{z - a} = 2\pi i.$$

First Proof: (Intuitive)

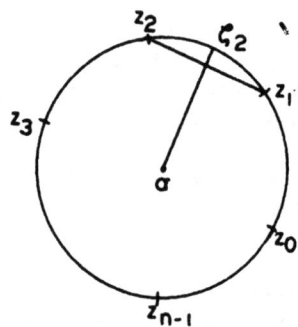

Let r be the radius of the circle. Let points of division $z_0, z_1, \ldots, z_n = z_0$ be taken on the circle. Choose ζ_j as the midpoint of the arc $\overline{z_{j-1} z_j}$.

Consider $\dfrac{1}{\zeta_j - a} (z_j - z_{j-1})$. Its modulus is $\dfrac{|z_j - z_{j-1}|}{|\zeta_j - a|}$ which, since ζ_j is on the circle, equals $\dfrac{|z_j - z_{j-1}|}{r}$.

The amplitude of $z_j - z_{j-1}$ is the inclination of the segment $\overline{z_{j-1} z_j}$. The amplitude of $\zeta_j - a$ is the inclination of the radius perpendicular to the segment $\overline{z_{j-1} z_j}$ and is less by $90°$ than the inclination of $\overline{z_{j-1} z_j}$.

Hence, the amplitude of $\dfrac{z_j - z_{j-1}}{\zeta_j - a}$ is $90°$. That is, $\dfrac{z_j - z_{j-1}}{\zeta_j - a}$ is a pure imaginary and we have

$$\frac{z_j - z_{j-1}}{\zeta_j - a} = \left|\frac{z_j - z_{j-1}}{r}\right| i.$$

104

Hence, an approximation to $\int_C \frac{dz}{z-a}$ is given by

$$\frac{i}{r}[|z_1 - z_0| + \ldots + |z_n - z_{n-1}|].$$

The quantity in brackets is the length of a polygon inscribed in the circle. Its limit is $2\pi r$, the length of the circle.

Hence,

$$\int_C \frac{dz}{z-a} = 2\pi i.$$

Second Proof: (Arithmetic)

We assume $a = 0$. The proof can easily be generalized to take care of all values of a. We shall show that the integral over that portion of the positively sensed circle which lies in the first quadrant is $\frac{\pi i}{2}$. Each of the other three quadrants would similarly be found to give an equal result. Let

$$x = \varphi(t), \quad y = \psi(t) \quad c \leq t \leq d$$

represent that portion of the circle which lies in the first quadrant. Let

$$t_0 = c, t_1, t_2, \ldots, t_n = d$$

be chosen on (c, d) and let

$$z_0 = r, z_1, z_2, \ldots, z_n = ri$$

be the corresponding points on the circle.

Consider any arc $\widehat{z_{j-1} z_j}$. We shall prove that there is a point ζ_j on it, that is, a point corresponding to some value of t between t_{j-1} and t_j, such that

$$\frac{z_j - z_{j-1}}{\zeta_j}$$

is a pure imaginary, and, indeed, the product of i by a positive number.

Let

$$z_j = x_j + iy_j, \quad z_{j-1} = x_{j-1} + iy_{j-1}.$$

Then

$$x_j < x_{j-1}, \quad y_j > y_{j-1}.$$

Consider the straight line

(1) $$y = \frac{y_{j-1} + y_j}{x_{j-1} + x_j} x.$$

The points of our circle are those points $x + yi$ for which

(2) $$x^2 + y^2 = r^2.$$

One sees from the elements of algebra that the straight line and circle have two points in common, one of which lies in the first quadrant and the other in the third.

Let ζ_j be the intersection in the first quadrant. We shall prove that ζ_j lies on $\widehat{z_{j-1} z_j}$.

For this, it will suffice to show that the abscissa of ζ_j lies between x_j and x_{j-1}. For then the t of ζ_j will be known to lie between t_{j-1} and t_j.

We have, since
$$y_j > y_{j-1}, \quad x_j < x_{j-1},$$
$$\frac{y_{j-1} + y_j}{x_{j-1} + x_j} > \frac{y_{j-1}}{x_{j-1}}$$

Hence, for the x and y of ζ_j, we have, by (1),
$$y > \frac{y_{j-1}}{x_{j-1}} x.$$

Then, by (2),
$$x^2 \left(1 + \frac{y_{j-1}^2}{x_{j-1}^2}\right) < r^2.$$

Since $x_{j-1}^2 + y_{j-1}^2 = r^2$, we have
$$x^2 \frac{r^2}{x_{j-1}^2} < r^2,$$

so that
$$x^2 < x_{j-1}^2$$

and
$$x < x_{j-1}.$$

Similarly,
$$x > x_j.$$

We shall prove now that
$$\frac{z_j - z_{j-1}}{\zeta_j}$$

is the product of i by a positive number.

Since, by (1), the y and x of ζ_j are proportional to $y_{j-1} + y_j$ and $x_{j-1} + x_j$, and since the four quantities just mentioned are positive, it will suffice to prove that
$$\frac{z_j - z_{j-1}}{x_{j-1} + x_j + (y_{j-1} + y_j) i}$$

or, what is the same

(3)
$$\frac{x_j - x_{j-1} + (y_j - y_{j-1}) i}{x_{j-1} + x_j + (y_{j-1} + y_j) i}$$

is the product of i by a positive number.

Now
$$x_j^2 + y_j^2 = x_{j-1}^2 + y_{j-1}^2 = r^2$$

so that

$$x_j^2 - x_{j-1}^2 = y_{j-1}^2 - y_j^2$$

or

(4) $$(x_j + x_{j-1})(x_j - x_{j-1}) = -(y_j + y_{j-1})(y_j - y_{j-1}).$$

We multiply numerator and denominator of (3) by $x_j - x_{j-1}$ and find (3) to equal

$$(x_j - x_{j-1}) \frac{(x_j - x_{j-1}) + (y_j - y_{j-1})i}{(x_j + x_{j-1})(x_j - x_{j-1}) + (y_j + y_{j-1})(x_j - x_{j-1})i}$$

or, by (4),

$$(x_j - x_{j-1}) \frac{(x_j - x_{j-1}) + (y_j - y_{j-1})i}{-(y_j + y_{j-1})(y_j - y_{j-1}) + (y_j + y_{j-1})(x_j - x_{j-1})i}$$

or

$$\left[\frac{x_j - x_{j-1}}{y_j + y_{j-1}}\right] \frac{(x_j - x_{j-1}) + (y_j - y_{j-1})i}{-(y_j - y_{j-1}) + (x_j - x_{j-1})i}$$

or

$$-\frac{x_j - x_{j-1}}{y_j + y_{j-1}} i,$$

which is the product of i by a positive number.

We have thus

$$\frac{z_j - z_{j-1}}{\zeta_j} = \frac{|z_j - z_{j-1}|}{r} i$$

and the discussion continues precisely as in the intuitive proof.

XXV

The Cauchy Integral Formula and the Derivatives of an Analytic Function

CAUCHY INTEGRAL FORMULA

1. Let A be an open region. Let $f(z)$ be a function analytic throughout A. Consider an open region B in A, bounded by one or more rectifiable curves which lie in A. We shall prove the Cauchy Integral Formula, which states that, for any z in B, we have

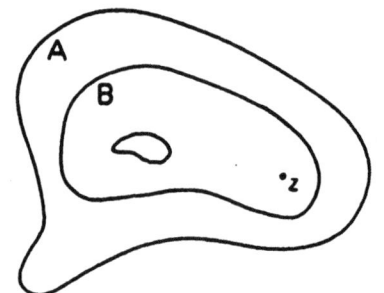

$$f(z) = \frac{1}{2\pi i} \int_C \frac{f(\zeta)\, d\zeta}{\zeta - z},$$

where $\int_C$ denotes the integral around the complete contour of B in the positive sense; that is the sum of the integral around the outer contour in the positive sense and of the integrals around the inner contours in the negative sense.

Proof: Let z be any fixed point of B. Let Γ be a circle, with z as center, lying, with its interior, in B. The expression

$$\frac{f(\zeta)}{\zeta - z}$$

defines a function of ζ which is analytic in the open region obtained by deleting z from A. Hence the integral of this function around the complete contour of B in the positive sense, added to the integral around Γ in the negative sense, will give zero. That is,

$$(1) \qquad \int_\Gamma \frac{f(\zeta)\, d\zeta}{\zeta - z} = \int_C \frac{f(\zeta)\, d\zeta}{\zeta - z}.$$

In the first member of (1), the integration is around Γ in the positive sense.

Now

$$(2) \qquad \int_\Gamma \frac{f(\zeta)\, d\zeta}{\zeta - z} = \int_\Gamma \frac{f(z)\, d\zeta}{\zeta - z} + \int_\Gamma \frac{f(\zeta) - f(z)}{\zeta - z}\, d\zeta = 2\pi i\, f(z) + \int_\Gamma \frac{f(\zeta) - f(z)}{\zeta - z}\, d\zeta.$$

Let any $\varepsilon > 0$ be preassigned. Let $\delta > 0$ be such that $|f(\zeta) - f(z)| < \varepsilon$ for $|\zeta - z| < \delta$. The existence of δ follows from the continuity of $f(z)$. Let the radius of Γ be taken as less than δ and let r represent that radius. Then, for any ζ on Γ,

$$|\zeta - z| = r.$$

Also, the length of Γ is $2\pi r$.

Hence

$$\left| \int_\Gamma \frac{f(\zeta) - f(z)}{\zeta - z}\, d\zeta \right| \leq \frac{\varepsilon}{r} 2\pi r = 2\pi \varepsilon.$$

Then, by (2)

$$\left| f(z) - \frac{1}{2\pi i} \int_\Gamma \frac{f(\zeta)\, d\zeta}{\zeta - z} \right| \leq \varepsilon,$$

so that, by (1),

$$\left|f(z) - \frac{1}{2\pi i} \int_C \frac{f(\zeta)\, d\zeta}{\zeta - z}\right| \le \epsilon.$$

As ϵ is arbitrarily small, we have

$$f(z) = \frac{1}{2\pi i} \int_C \frac{f(\zeta)\, d\zeta}{\zeta - z}.$$

Q.E.D.

THE FIRST DERIVATIVE

2. We shall prove that, for any z of B,

$$f'(z) = \frac{1}{2\pi i} \int_C \frac{f(\zeta)\, d\zeta}{(\zeta - z)^2}.$$

Proof: We deal with a fixed z in B. Let $r > 0$ be such that z is the center of a circle of radius $2r$ which lies, with its interior, in B. Then, for every ζ of C,

(3)
$$|\zeta - z| > 2r$$

and, if $|\Delta z| < r$,

(4)
$$|\zeta - (z + \Delta z)| > r.$$

We have, for $|\Delta z| < r$,

$$f(z + \Delta z) - f(z) = \frac{1}{2\pi i} \int_C f(\zeta) \left[\frac{1}{\zeta - (z + \Delta z)} - \frac{1}{\zeta - z}\right] d\zeta$$

$$= \frac{1}{2\pi i} \int_C \frac{\Delta z\, f(\zeta)}{[\zeta - (z + \Delta z)](\zeta - z)}\, d\zeta.$$

Hence

$$\frac{f(z + \Delta z) - f(z)}{\Delta z} = \frac{1}{2\pi i} \int_C \frac{f(\zeta)\, d\zeta}{[\zeta - (z + \Delta z)](\zeta - z)}$$

Thus

(5)
$$\frac{f(z + \Delta z) - f(z)}{\Delta z} - \frac{1}{2\pi i} \int_C \frac{f(\zeta)\, d\zeta}{(\zeta - z)^2}$$

$$= \frac{1}{2\pi i} \int_C f(\zeta) \left[\frac{1}{(\zeta - z - \Delta z)(\zeta - z)} - \frac{1}{(\zeta - z)^2}\right] d\zeta$$

$$= \frac{\Delta z}{2\pi i} \int_C \frac{f(\zeta)\, d\zeta}{(\zeta - z)^2 (\zeta - z - \Delta z)}.$$

As $f(z)$ is integrable along C, $f(z)$ is bounded on C. Let $M > 0$ be such that

$$|f(\zeta)| < M$$

for every ζ on C. Then, by (3), (4) and (5), if λ is the sum of the lengths of the contours which make up C,

$$\left|\frac{f(z + \Delta z) - f(z)}{\Delta z} - \frac{1}{2\pi i} \int_C \frac{f(\zeta)\, d\zeta}{(\zeta - z)^2}\right| \le \frac{|\Delta z|}{2\pi} \frac{M}{4r^3} \lambda.$$

As the quantity $\frac{M\lambda}{8\pi r^3}$ is fixed, we see that

$$f'(z) = \frac{1}{2\pi i} \int_C \frac{f(\zeta)\, d\zeta}{(\zeta - z)^2}.$$

Q.E.D.

THE HIGHER DERIVATIVES OF AN ANALYTIC FUNCTION

3. Let $f(z)$ be analytic in an open region A. We know that $f(z)$ has a derivative, $f'(z)$, in A. We are going to prove that $f'(z)$ is analytic in A. It will be an immediate consequence of that that $f(z)$ *possesses derivatives of all orders in A, which are all analytic functions.*

We see thus a striking difference between the real domain and the complex. A function of a real variable can have a first derivative without having a second derivative.

The result is established in the following form. Using the notation of the preceding sections, one shows that, for every positive integer n and for every z of B, $f(z)$ has a derivative of order n, given by the formula

$$f^{(n)}(z) = \frac{n!}{2\pi i} \int_C \frac{f(\zeta)\, d\zeta}{(\zeta - z)^{n+1}}$$

We shall carry out the details of the proof for $n = 2$. Let r be defined as in the preceding section. We have, for $|\Delta z| < r$,

$$f'(z + \Delta z) - f'(z) = \frac{1}{2\pi i} \int_C f(\zeta) \left[\frac{1}{(\zeta - z - \Delta z)^2} - \frac{1}{(\zeta - z)^2} \right] d\zeta$$

$$= \frac{\Delta z}{2\pi i} \int_C f(\zeta) \frac{2(\zeta - z) - \Delta z}{(\zeta - z - \Delta z)^2 (\zeta - z)^2} d\zeta$$

$$= \frac{\Delta z}{2\pi i} \int_C f(\zeta) \left[\frac{2}{(\zeta - z - \Delta z)^2 (\zeta - z)} - \frac{\Delta z}{(\zeta - z - \Delta z)^2 (\zeta - z)^2} \right] d\zeta.$$

Hence

$$\frac{f'(z + \Delta z) - f'(z)}{\Delta z} - \frac{2!}{2\pi i} \int_C \frac{f(\zeta)\, d\zeta}{(\zeta - z)^3} =$$

$$= \frac{1}{2\pi i} \int_C f(\zeta) \left[\frac{2}{(\zeta - z - \Delta z)^2 (\zeta - z)} - \frac{2}{(\zeta - z)^3} - \frac{\Delta z}{(\zeta - z - \Delta z)^2 (\zeta - z)^2} \right] d\zeta$$

$$= \frac{1}{2\pi i} \int_C f(\zeta) \left[\frac{4(\zeta - z)\Delta z - 2\Delta z^2 - (\zeta - z)\Delta z}{(\zeta - z - \Delta z)^2 (\zeta - z)^3} \right] d\zeta$$

$$= \frac{\Delta z}{2\pi i} \int_C f(\zeta) \frac{3(\zeta - z) - 2\Delta z}{(\zeta - z - \Delta z)^2 (\zeta - z)^3} d\zeta$$

Let $G > 0$ be such that, for $|\Delta z| < r$ and for ζ on C,

$$|3(\zeta - z) - 2\Delta z| < G.$$

Then, if λ is the sum of the lengths of the contours which make up C, and M an upper bound for $|f(\zeta)|$ on C,

$$\left| \frac{f'(z + \Delta z) - f'(z)}{\Delta z} - \frac{2!}{2\pi i} \int_C \frac{f(\zeta)\, d\zeta}{(\zeta - z)^3} \right| \leq \frac{MG\lambda\, |\Delta z|}{2\pi r^2 \cdot 8r^2}$$

so that, when Δz approaches zero,

$$\frac{f'(z + \Delta z) - f'(z)}{\Delta z}$$

approaches a limit equal to

$$\frac{2!}{2\pi i} \int_C \frac{f(\zeta)\, d\zeta}{(\zeta - z)^3}.$$

Hence $f''(z)$ exists and is equal to

$$\frac{2!}{2\pi i} \int_C \frac{f(\zeta)\, d\zeta}{(\zeta - z)^3}.$$

4. On the basis of §2, we can formulate the following theorem:

Let $F(z)$ be defined, and continuous, at every point of the complete contour C of the open region B. (C and B as before.) Then

$$\int_C \frac{F(\zeta)\, d\zeta}{\zeta - z}$$

represents a function analytic throughout B. Its nth derivative is given by

$$n! \int_C \frac{F(\zeta)\, d\zeta}{(\zeta - z)^{n+1}}.$$

XXVI

Infinite Sequences and Infinite Series of Analytic Functions

UNIFORM CONVERGENCE

1. Let

$$s_1(z), s_2(z), \ldots, s_n(z), \ldots$$

be a sequence of functions of z (not necessarily analytic), which converges on a domain E to a limit $s(z)$. We shall say that the sequence converges *uniformly* on E if, for every $\varepsilon > 0$ an N can be found such that, for $n > N$, we have

$$|s(z) - s_n(z)| < \varepsilon$$

for every z on E.

As in the theory of the real variable, one proves easily that, for the sequence to converge uniformly on E, it is necessary and sufficient that for every $\varepsilon > 0$ an N exist such that, for every $n > N$ and for every p,

$$|s_{n+p}(z) - s_n(z)| < \varepsilon$$

for every z on E.

CONTINUITY AND INTEGRABILITY

2. Let C represent a simple curve or a simple closed curve. We take these special types of curves because they are the only types which we shall have to use. What follows applies really even to curves which intersect themselves.

Let

$$s_1(z), s_2(z), \ldots, s_n(z), \ldots$$

be a sequence of functions, each of which is defined and continuous along C. Let the sequence converge *uniformly* on C to a limit $s(z)$. We have then the theorem:

Theorem: The limit $s(z)$ is continuous at every point of C.

Proof: As for the real variable.

Now, let C be rectifiable.

Since $s(z)$ is continuous along C, it is *integrable* along C.

We say that the *sequence*

$$\int_C s_1(z)\, dz, \ldots, \int_C s_n(z)\, dz, \ldots$$

converges and that its limit is $\int_C s(z)\, dz$.

Proof: Let the length of C be λ. An $\varepsilon > 0$ being given, suppose that, for $n > N$, $|s(z) - s_n(z)| < \varepsilon$ all along C. Then

$$\left|\int_C s(z)\, dz - \int_C s_n(z)\, dz\right| = \left|\int_C [s(z) - s_n(z)]\, dz\right| \leq \varepsilon\lambda.$$

Q.E.D.

3. Let A be an open region. Let each function of the sequence

(1) $$s_1(z), \ldots, s_n(z), \ldots$$

be analytic in A. Suppose that (1) converges uniformly to a limit $s(z)$ throughout A.

We say that $s(z)$ *is analytic in* A.

Proof: Let z_0 be any point of A. We shall show that $s(z)$ is analytic at z_0. Let Γ be a circle with z_0 as center which lies, with its interior, in A. For any particular z within Γ, the sequence

$$\frac{s_1(\zeta)}{\zeta - z}, \ldots \frac{s_n(\zeta)}{\zeta - z}, \ldots$$

converges for every ζ on Γ to $\frac{s(\zeta)}{\zeta - z}$. It is easy to prove that the convergence is uniform along Γ. Hence the sequence

(2) $$\frac{1}{2\pi i} \oint_\Gamma \frac{s_1(\zeta)}{\zeta - z} d\zeta, \ldots, \frac{1}{2\pi i} \oint_\Gamma \frac{s_n(\zeta)}{\zeta - z} d\zeta, \ldots$$

converges to

$$\frac{1}{2\pi i} \oint_\Gamma \frac{s(\zeta) \, d\zeta}{\zeta - z}.$$

But, for z, within Γ, the sequence (2) is identical with (1). Hence, for z within Γ,

$$s(z) = \frac{1}{2\pi i} \oint_\Gamma \frac{s(\zeta)}{\zeta - z} d\zeta.$$

But the last integral represents a function analytic within Γ. Hence $s(z)$ is analytic within Γ.

Q.E.D.

We shall show that the *sequence of derivatives*

(3) $$s_1'(z), \ldots, s_n'(z), \ldots$$

converges to $s'(z)$ throughout A.

We have, for any z in Γ,

$$s'(z) = \frac{1}{2\pi i} \oint_\Gamma \frac{s(\zeta)}{(\zeta - z)^2} d\zeta.$$

Now the integral $\oint_\Gamma \frac{s(\zeta) \, d\zeta}{(\zeta - z)^2}$ is the limit of the sequence

(4) $$\oint_\Gamma \frac{s_1(\zeta)}{(\zeta - z)^2} d\zeta, \ldots, \oint_\Gamma \frac{s_n(\zeta)}{(\zeta - z)^2} d\zeta, \ldots,$$

because of the uniformity of the convergence along Γ of the sequence

$$\frac{s_1(\zeta)}{(\zeta - z)^2}, \ldots, \frac{s_n(\zeta)}{(\zeta - z)^2}, \ldots$$

for any particular z within Γ. As the nth term of (4) is $2\pi i \, s_n'(z)$, we see that (3) converges to $s'(z)$ within Γ.

We shall prove that, on any closed and bounded domain interior to A, *the sequence (3) converges uniformly.*

Proof: Let Γ be as above. Let $2r$ be its radius. Let Γ' be a circle with z as center and r as radius. Choosing an $\varepsilon > 0$, suppose that, for $n > N$, we have $|s(\zeta) - s_n(\zeta)| < \varepsilon$ along Γ. For ζ on Γ and for z within Γ', we have $|\zeta - z| > r$. Hence, for z in Γ' and $n > N$,

$$|s'(z) - s_n'(z)| = \frac{1}{2\pi} |\int_\Gamma \frac{s(\zeta) - s_n(\zeta)}{(\zeta - z)^2} d\zeta| \leq \frac{1}{2\pi} \frac{\varepsilon}{r^2} 2\pi (2r).$$

[The length of Γ is $2\pi (2r)$.]

Thus, (3) converges uniformly in Γ'. Borel's theorem does the rest.

THE HIGHER DERIVATIVES

4. Exactly as in §3, we prove that if

(4) $\qquad s_1(z), \ldots, s_n(z), \ldots$

is a sequence of functions analytic in an open region A, the sequence converging uniformly in A to a function $s(z)$ (necessarily analytic), then, if p is any positive integer, the sequence of the pth derivatives of the functions in (1) converges to the pth derivative of $s(z)$ throughout A, the convergence being uniform in every closed and bounded domain contained in A. One uses, in the proof, the formula

$$s_n^{(p)}(z) = \frac{p!}{2\pi i} \int_\Gamma \frac{s_n(\zeta) \, d\zeta}{(\zeta - z)^{p+1}}.$$

SERIES OF COMPLEX NUMBERS

5. If

$$a_1, a_2, \ldots, a_n, \ldots$$

is an infinite sequence of complex numbers, we mean by the *sum* of the *series*

(5) $\qquad a_1 + a_2 + \ldots + a_n + \ldots$

the limit, as n increase indefinitely, of

$$s_n = a_1 + \ldots + a_n,$$

provided that such a limit exists. When the limit exists, we say that (5) is *convergent*. As in the real variable, we can prove that (5) converges if

(6) $\qquad |a_1| + |a_2| + \ldots + |a_n| + \ldots$

converges. If (6) converges, we say that (5) is *absolutely* convergent.

SERIES OF ANALYTIC FUNCTIONS

6. Let

$$u_1(z), u_2(z), \ldots, u_n(z), \ldots$$

be a sequence of functions defined on a domain E. Suppose that, for every (z) on E, the series

(7) $\qquad u_1(z) + \ldots + u_n(z) + \ldots$

converges. We then use the expression -- *The infinite series (7) of functions converges throughout E.*

We represent by $s_n(z)$ the sum

$$u_1(z) + \ldots + u_n(z).$$

If (7) converges throughout E to the sum $s(z)$, we call

$$s(z) - s_n(z) = u_{n+1}(z) + u_{n+2}(z) + \ldots$$

the remainder after n *terms of (7).*

Let (7) converge on E and let $R_n(z)$ denote its remainder after n terms. We call (7) *uniformly convergent on* E if, for every $\varepsilon > 0$, an N exists such that, for $n > N$,

$$|R_n(z)| < \varepsilon$$

throughout E.

The theorems on sequences of analytic functions go over immediately into theorems on series. Thus, *let*

$$u_1(z) + \ldots + u_n(z) + \ldots,$$

each $u_n(z)$ being analytic in an open region A, converge uniformly throughout A to a sum $s(z)$. Then $s(z)$ is analytic throughout A. Furthermore, for every z in A,

$$s'(z) = u_1'(z) + \ldots + u_n'(z) + \ldots,$$

and the series in the second member of the last equation is uniformly convergent in every closed and bounded domain contained in A.

XXVII

Power Series

GREATEST LIMIT OF A SEQUENCE OF REAL NUMBERS

1. Consider a sequence of non-negative real numbers

$$(1) \qquad a_1, a_2, \ldots, a_n, \ldots$$

which is bounded from above, that is, a sequence for which a number M exists such that

$$a_n < M$$

for every n.

We shall prove the existence of a number l which has the following two properties:

(α) *For every $\varepsilon > 0$, there are an infinite number of elements of the sequence which exceed $l - \varepsilon$.*

(β) *For every $\varepsilon > 0$, there are at most a finite number of elements which exceed $l + \varepsilon$.*

Proof: Let B represent the totality of numbers which have the property of being exceeded by an infinite number of elements of the sequence. (Evidently all negative numbers belong to B.) Then B is bounded from above, since M exceeds every number in B. Let l be the least upper bound of B. Then l has the properties (α) and (β).

Evidently no number distinct from l as just determined has the properties (α) and (β).

We shall call l the *greatest limit* of the sequence.

For a sequence (1) which is not bounded from above, we shall use the expression -- *The greatest limit of the sequence is plus infinity.* The statement just made is to be regarded merely as a convenient substitute for the statement that the sequence is unbounded. It does not at all imply that we have created a number "plus infinity."

POWER SERIES

2. We consider a series

$$(2) \qquad c_0 + c_1(z-a) + c_2(z-a)^2 + \ldots + c_n(z-a)^n + \ldots$$

where the c's and $\underline{a}$ are constants (complex numbers). We shall call (2) a *power series*. We are going to study the values of z for which the series converges.

Consider the sequence

$$|c_0|, |c_1|, \sqrt{|c_2|}, \sqrt[3]{|c_3|}, \ldots, \sqrt[n]{|c_n|}, \ldots$$

By $\sqrt[n]{|c_n|}$, we mean the unique non-negative number whose nth power equals $|c_n|$. Let l be the greatest limit of this sequence.

We shall show that:

(α) If l is infinite, the power series diverges for every z except $z = a$.

(β) If $l > 0$ and if l is not infinite, the series converges for $|z - a| < \frac{1}{l}$ and diverges for $|z - a| > \frac{1}{l}$.

116

(γ) If $l = 0$, the series converges for every value of z.

The results just stated are due to Cauchy and to Hadamard.

We shall work under the assumption that $a = 0$. The extension to the general case is immediate. Thus, we consider the series

$$(3) \qquad c_0 + c_1 z + c_2 z^2 + \ldots + c_n z^n + \ldots$$

Let us dispose of case (α).

Take any z distinct from zero.

As l is infinite, an infinite number of the quantities $\sqrt[n]{|c_n|}$ exceed $\frac{1}{|z|}$. Hence, for an infinite number of values of n,

$$\sqrt[n]{|c_n|} \cdot |z| > 1$$

or

$$|c_n z^n| > 1.$$

Thus, the nth term of (3) does not approach 0, and (3) diverges.

We take now case (β).

Let $r = \frac{1}{l}$. We shall prove first that *(3) diverges for $|z| > r$ and then that, if r_1 is any positive number less than r, (3) converges absolutely and uniformly for $|z| < r_1$*. This will certainly establish our statement above.

Consider any z with $|z| > r$. Then $\frac{1}{|z|} < l$, so that, for an infinite number of values of n,

$$\sqrt[n]{|c_n|} > \frac{1}{|z|}$$

or

$$|c_n z^n| > 1.$$

This, as above, implies divergence.

Now, r_1 being as above, let r_2 be any number such that

$$r_1 < r_2 < r.$$

Then $\frac{1}{r_2} > l$, so that only a finite number of the quantities $\sqrt[n]{|c_n|}$ exceed $\frac{1}{r_2}$. Hence, there is an N such that, for $n > N$,

$$\sqrt[n]{|c_n|} \leq \frac{1}{r_2}$$

or

$$|c_n r_2^n| \leq 1.$$

For $|z| < r_1$, each term of the series (3) has a modulus not greater than the corresponding term of

$$(4) \qquad |c_0| + |c_1| r_1 + \ldots + |c_n| r_1^n + \ldots$$

Hence, if we can show that (4) converges, we shall know that (3) converges absolutely and uniformly for $|z| \leq r_1$.

We write (4)

$$(5) \qquad |c_0| + |c_1 r_2| \frac{r_1}{r_2} + |c_2 r_2^2| \left(\frac{r_1}{r_2}\right)^2 + \ldots + |c_n r_2^n| \left(\frac{r_1}{r_2}\right)^n + \ldots$$

Since, for n large, $|c_n r_2^n| < 1$, the terms of (5) are ultimately not greater than those of

(6) $$1 + \frac{r_1}{r_2} + \left(\frac{r_1}{r_2}\right)^2 + \ldots + \left(\frac{r_1}{r_2}\right)^n + \ldots$$

As (6) is a geometric series with ratio less than unity in absolute value, (6) is convergent. Hence (5) is convergent and our result is proved.

Of course, (3) converges *absolutely* for every z such that $|\dot{z}| < r$.

We consider finally the case (γ) and prove that, for every $r_1 > 0$, (3) converges absolutely and uniformly for $|z| < r_1$.

Proof: Let $r_2 > r_1$. As $\frac{1}{r_2} > 0$, only a finite number of the quantities $\sqrt[n]{|c_n|}$ exceed $\frac{1}{r_2}$. The proof continues precisely as for case (β).

CIRCLE OF CONVERGENCE

3. When $0 < l < +\infty$, we call $r = \frac{1}{l}$ *the radius of convergence* of the series

(7) $$a_0 + c_1(z-a) + \ldots + c_n(z-a)^n + \ldots$$

We call the circle $|z - a| = r$ the *circle of convergence* of (7). When $l = 0$, we use the expression "the radius of convergence is infinite." In that case, we call the entire plane "the interior of the circle of convergence."

Thus, inside of its circle of convergence, (7) converges absolutely at every point. Outside the circle of convergence (7) diverges. Upon the circle of convergence, various states of affairs may exist. The series may converge all over the circle, or may diverge everywhere on it, or be convergent at some points and divergent at others.

Within any circle interior to the circle of convergence, the series converges uniformly. Thus, *the series represents a function which is analytic in the interior of the circle of convergence*. According to our results on series of analytic functions, the derivative of the function represented by (7) is obtained by differentiating the series formally.

Examples:

(A) $$1 + z + z^2 + \ldots + z^n + \ldots$$

$r = 1$. Series divergent for $|z| = 1$.

(B) $$\frac{z}{1^2} + \frac{z^2}{2^2} + \ldots + \frac{z^n}{n^2} + \ldots$$

$r = 1$. Convergent for $|z| = 1$.

(C) $$1 + z + \frac{z^2}{2!} + \ldots + \frac{z^n}{n!} + \ldots$$

$r = \infty$.

(D) $$1 + 1!z + 2!z^2 + \ldots + n!z^n + \ldots$$

$r = 0$.

(E) $$z - \frac{z^2}{2} + \frac{z^3}{3} - \frac{z^4}{4} + \frac{z^5}{5} - \ldots$$

$r = 1$. Divergence on circle of convergence at $z = -1$; convergence everywhere else on circle. The last statement is difficult to prove except for $z = 1$.

XXVIII

Taylor's Expansion

STATEMENT OF RESULTS

1. Let $f(z)$ be analytic in an open region A. Let <u>a</u> be any point of A and let Γ be any circle with <u>a</u> as center, lying with its interior, in A. We shall prove that $f(z)$ admits, throughout the interior of Γ, a representation.

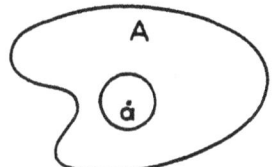

$$f(z) = c_0 + c_1(z-a) + c_2(z-a)^2 + \ldots + c_n(z-a)^n + \ldots$$

We shall prove that there is but one power series of this type (that is, a series of powers of $z - a$), which represents $f(z)$ for a neighborhood of <u>a</u> and that its coefficients are given by

$$c_0 = f(a), \quad c_n = \frac{f^{(n)}(a)}{n!}, \quad (n > 0).$$

Remark: Actually the power series will converge to $f(z)$ upon Γ, as well as within Γ, for it is easy to show that a circle concentric with Γ and of radius only slightly greater than Γ, lies with its interior in A.

PROOF OF UNIQUENESS

2. Suppose that the series

$$c_0 + c_1(z-a) + c_2(z-a)^2 + \ldots + c_n(z-a)^n + \ldots$$

converges to $f(z)$ for a neighborhood of <u>a</u>. We find, putting $z = a$, that $f(a) = c_0$. For a sufficiently small neighborhood of <u>a</u>, the power series will converge uniformly, so that it can be differentiated term by term. Thus

$$f'(z) = c_1 + 2c_2(z-a) + \ldots + nc_n(z-a)^{n-1} + \ldots$$

Hence, putting $z = a$, we find

$$f'(a) = c_1.$$

Again, we have

$$f''(z) = 2c_2 + 3 \cdot 2 c_3(z-a) + \ldots$$

so that $f''(a) = 2c_2$.

Proceeding in this way, we see that, for $n \geq 1$,

$$c_n = \frac{f^{(n)}(a)}{n!}.$$

Hence, there can be no more than one expansion of $f(z)$ in powers of $(z - a)$, valid for a neighborhood of <u>a</u>.

A PRELIMINARY

3. We shall need the identity, easily proved,

$$\frac{1}{A-B} = \frac{1}{A} + \frac{B}{A^2} + \frac{B^2}{A^3} + \ldots + \frac{B^n}{A^{n+1}} + \frac{B^{n+1}}{A^{n+1}(A-B)}$$

Here, it is understood that A and B are complex numbers with $A \neq 0$ and $A - B \neq 0$

DERIVATION OF EXPANSION

4. We refer to §1. We have, for z within Γ,

$$f(z) = \frac{1}{2\pi i} \int_\Gamma \frac{f(\zeta)\, d\zeta}{\zeta - z},$$

or

$$f(z) = \frac{1}{2\pi i} \int_\Gamma \frac{f(\zeta)\, d\zeta}{(\zeta - a) - (z - a)}.$$

Now

$$\frac{1}{(\zeta - a) - (z - a)} = \frac{1}{\zeta - a} + \frac{z - a}{(\zeta - a)^2} + \ldots + \frac{(z - a)^n}{(\zeta - a)^{n+1}} + \frac{(z - a)^{n+1}}{(\zeta - a)^{n+1}(\zeta - z)}$$

Hence

$$f(z) = \frac{1}{2\pi i} \int_\Gamma \frac{f(\zeta)\, d\zeta}{\zeta - a} + \frac{(z - a)}{2\pi i} \int_\Gamma \frac{f(\zeta)\, d\zeta}{(\zeta - a)^2} + \ldots + \frac{(z - a)^n}{2\pi i} \int_\Gamma \frac{f(\zeta)\, d\zeta}{(\zeta - a)^{n+1}}$$
$$+ \frac{(z - a)^{n+1}}{2\pi i} \int_\Gamma \frac{f(\zeta)\, d\zeta}{(\zeta - a)^{n+1}(\zeta - z)}.$$

Now

$$\frac{1}{2\pi i} \int_\Gamma \frac{f(\zeta)\, d\zeta}{\zeta - a} = f(a)$$

and

$$\frac{1}{2\pi i} \int_\Gamma \frac{f(\zeta)\, d\zeta}{(\zeta - a)^{m+1}} = \frac{f^{(m)}(a)}{m!}$$

for $m > 0$.

Hence, for z within Γ,

$$f(z) = f(a) + f'(a)(z - a) + \ldots + \frac{f^{(n)}(a)}{n!}(z - a)^n + \frac{(z - a)^{n+1}}{2\pi i} \int_\Gamma \frac{f(\zeta)\, d\zeta}{(\zeta - a)^{n+1}(\zeta - z)}.$$

To prove the validity of the Taylor expansion, all we have to show is that as n increases,

$$(z - a)^{n+1} \int_\Gamma \frac{f(\zeta)\, d\zeta}{(\zeta - a)^{n+1}(\zeta - z)}$$

tends towards zero.

Let r be the radius of Γ. Then $|z - a| < r$. Also

$$|\zeta - z| = |(\zeta - a) - (z - a)| \geq r - |z - a|.$$

Let M be such that $|f(z)| \leq M$ on Γ. Such an M exists, since $f(z)$ is continuous on Γ. Then

$$\left| \frac{(z - a)^{n+1}}{2\pi i} \int_\Gamma \frac{f(\zeta)\, d\zeta}{(\zeta - a)^{n+1}(\zeta - z)} \right| \leq \frac{|z - a|^{n+1} M}{2\pi r^{n+1}[r - |z - a|]} 2\pi r = \frac{Mr}{r - |z - a|} \left(\frac{|z - a|}{r}\right)^{n+1}$$

Now $\frac{|z - a|}{r}$ is less than unity, so that its $n + 1$st power, if n is large, is very small. This settles the question of the convergence of the power series to $f(z)$.

Note: We have used the fact that, if $0 < h < 1$, then h^n approaches 0 as n increases indefinitely. This can be proved as follows:

Let $k = \frac{1}{h}$. Then $k > 1$. Now

$$k^n = [1 + (k - 1)]^n = 1 + n(k - 1) + \frac{n(n - 1)}{2!}(k - 1)^2 + \ldots$$

Since $k - 1 > 0$, we have $k^n > 1 + n(k - 1)$. Thus k^n is very large when n is large. Then $h^n = \frac{1}{k^n}$ is small when n is large.

XXIX

Liouville's Theorem and the Fundamental Theorem of Algebra

INTEGRAL FUNCTIONS AND LIOUVILLE'S THEOREM

1. A function which is analytic throughout the complex plane is called an *integral function*.

If $f(z)$ is an integral function and if a is any point of the complex plane, then $f(z)$ admits an expansion in powers of $z - a$ which is valid all over the plane.

In particular, taking $a = 0$, we have an expansion, valid for every z.

$$(1) \qquad f(z) = c_0 + c_1 z + \ldots + c_n z^n + \ldots$$

where $c_0 = f(0)$,

$$c_n = \frac{f^{(n)}(0)}{n!}, \qquad n = 1, 2, \ldots.$$

Thus, C being any circle with center at the origin, we have

$$(2) \qquad c_n = \frac{1}{2\pi i} \int_C \frac{f(\zeta)\, d\zeta}{\zeta^{n+1}}$$

We shall now prove Liouville's theorem, which is to the following effect:

An integral function which is bounded all over the plane is a constant.

Our hypothesis is:

(a) $f(z)$ is analytic for every value of z.
(b) M exists such that $|f(z)| < M$ for every z.

Our conclusion is: $f(z)$ is a constant.

Proof: In (2), let r be the radius of C. Then

$$(3) \qquad |c_n| \leq \frac{1}{2\pi} \frac{M}{r^{n+1}} 2\pi r \leq \frac{M}{r^n}.$$

Now (3) holds no matter how large r may be. If $n \geq 1$, r^n is very large when r is large. Thus, for $n \geq 1$,

$$c_n = 0$$

so that

$$f(z) = c_0$$

for every z.

Q.E.D.

THE FUNDAMENTAL THEOREM OF ALGEBRA

2. *Theorem: The equation*

$$a_0 z^n + a_1 z^{n-1} + \ldots + a_n = 0$$

where $n \geq 1$ and the a_i are constants with $a_0 \neq 0$, has at least one root.

Proof: Let $f(z)$ represent $a_0 z^n + \ldots + a_n$. Suppose that no value of z exists for which $f(z)$ is zero. We shall force a contradiction.

Our first step will be to prove that $|f(z)|$ is very large when $|z|$ is very large. We have

$$f(z) = z^n(a_0 + \frac{a_1}{z} + \ldots + \frac{a_n}{z^n}).$$

Now, when $|z|$ is very large, $|\frac{a_1}{z} + \ldots + \frac{a_n}{z^n}|$ is very small, so that

$$|a_0 + \frac{a_1}{z} + \ldots + \frac{a_n}{z^n}| > |\frac{a_0}{2}|.$$

Thus, for $|z|$ large,

$$|f(z)| > \frac{|a_0| \, |z|^n}{2}$$

and this proves our statement.

Let $r > 0$ be such that $|f(z)| > 1$ for $|z| > r$. Consider the function $\frac{1}{f(z)}$, which we shall represent by $g(z)$.

Since $f(z)$ vanishes nowhere, $g(z)$ is analytic everywhere, that is, $g(z)$ is an integral function.

Now for $|z| > r$, we have $|g(z)| < 1$.

On the other hand, $g(z)$ is certainly bounded for $|z| \leq r$. (Note that the domain for which $|z| \leq r$ is closed and bounded and that $g(z)$ is continuous at each point of this domain.)

Hence, $g(z)$ is bounded all over the plane, and, by Liouville's theorem, is a constant.

Thus $f(z)$, the reciprocal of $g(z)$, must be a constant. Hence $f(z)$ cannot be made arbitrarily large by taking $|z|$ large. This contradiction proves our theorem.

XXX

On the Zeros of Analytic Functions

THE CHIEF THEOREM

1. If $f(z)$ is analytic at $\underline{a}$ and if $f(a) = 0$, we shall call the point $\underline{a}$ a *zero* of $f(z)$.

Theorem: Let $f(z)$ be analytic throughout an open region A. Suppose that $f(z)$ is zero on a set of points of A which has a limit point in A. Then $f(z)$ is zero throughout A.

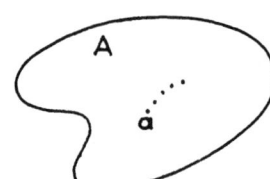

Proof: Let $\underline{a}$ be such a limit point as is mentioned in the theorem. We shall show first that $f(z)$ vanishes throughout some neighborhood of $\underline{a}$.

Let C be a circle, with $\underline{a}$ as center, lying with its interior in A. Then, within C, we have an expression for $f(z)$

$$f(z) = c_0 + c_1(z - a) + c_2(z - a)^2 + \ldots$$

We have $f(a) = c_0$. Now $f(z)$ is continuous at $\underline{a}$. Since $f(z)$ vanishes for points arbitrarily close to $\underline{a}$, we must have $c_0 = f(a) = 0$.

Thus, within C,

$$f(z) = c_1(z - a) + c_2(z - a)^2 + \ldots$$

For every z within C, except for $z = a$, the function

$$\frac{f(z)}{z - a}$$

is defined and is analytic. Also, if z is within C and $z \neq a$, we have

$$(1) \qquad \frac{f(z)}{z - a} = c_1 + c_2(z - a) + c_3(z - a)^2 + \ldots$$

Hence the power series in (1) converges at every point within C distinct from $\underline{a}$. By the nature of a power series, it must converge also for $z = a$. Thus the series represents an analytic function throughout the interior of C whose value at $\underline{a}$ is c_1. Hence, if we attribute to $\frac{f(z)}{z-a}$ the value c_1 at $\underline{a}$, $\frac{f(z)}{z-a}$ will be analytic throughout the interior of C. For $z \neq a$, $\frac{f(z)}{z-a}$ vanishes wherever $f(z)$ does. Hence $\frac{f(z)}{z-a}$ vanishes for points arbitrarily close to $\underline{a}$. Consequently, as $\frac{f(z)}{z-a}$ is continuous at $\underline{a}$, we have

$$\left. \frac{f(z)}{z - a} \right]_{z=a} = c_1 = 0.$$

We consider now the function

$$\frac{f(z)}{(z-a)^2} = c_2 + c_3(z - a) + \ldots$$

and reasoning as above, prove that $c_2 = 0$. Similarly, $c_n = 0$ for every n.

Thus $f(z)$ vanishes throughout the interior of C.

Consider now any point b of A. We shall prove that $f(b) = 0$.

Let $\underline{a}$ be joined to b by a continuous curve lying in A. This is possible, since A is an open region. Let such a curve be

$$z = \varphi(t) + i\psi(t) \qquad \alpha \leq t \leq \beta.$$

We have seen in previous work that there exists a $\delta > 0$ such that every point of the curve is the center of a circle of radius δ which lies, with its interior, in A.

Let $\eta > 0$ be such that, for $|t' - t| < \eta$, we have

$$|\varphi(t') + i\psi(t') - [\varphi(t) + i\psi(t)]| < \delta.$$

In (α, β), we take points

$$t_0 = \alpha,\ t_1 > t_0,\ t_2 > t_1,\ \ldots,\ t_n = \beta$$

with

$$t_{j+1} - t_j < \eta,\ j = 0,\ \ldots,\ n - 1.$$

Let the point of the curve which corresponds to t_j be z_j. Then

$$|z_{j+1} - z_j| < \delta,\ j = 0,\ \ldots,\ n - 1.$$

Consider z_1. It lies within a circle about $z = a$ of radius δ. But $f(z)$ vanishes throughout that circle. As z_1 is a limit point of points interior to the circle, $f(z)$ vanishes throughout any circle with center at z_1 which lies, with its interior, in A. Then $f(z)$ vanishes throughout a circle with center at z_1 and radius δ. This circle contains z_2. We continue in this fashion and find, finally, that $f(z_n) = f(b) = 0$.

Q.E.D.

SOME CONSEQUENCES OF THE PRECEDING THEOREM

2. What precedes shows that if $f(z)$ vanishes throughout an arbitrarily small area in A, it vanishes throughout A. Again, if $f(z)$ vanishes along a curve consisting of more than one point, it vanishes throughout A.

However, the theorem does not imply that $f(z)$ cannot vanish for an infinite set of points of A without vanishing identically. All that we know is that if such a set of points exists and if $f(z)$ does not vanish identically, then the set of points cannot have a limit point interior to A.

Suppose that $f(z)$ and $g(z)$ are both analytic throughout A and that $f(z) = g(z)$ for an infinite set of points of A with a limit point in A. Then $f(z) - g(z)$ is analytic throughout A, and, by what goes before, vanishes throughout A. Hence $f(z) = g(z)$ throughout A.

We shall formulate now a theorem known as the "Principle of the Permanence of Functional Equations."

Let A be an open region. Let $f_1(z),\ \ldots,\ f_n(z)$ be n functions, each analytic throughout A. Let

$$P(u_1,\ \ldots,\ u_n)$$

be a polynomial in $u_1,\ \ldots,\ u_n$ with constant coefficients. Suppose that for some area in A,

(2) $$P[f_1(z),\ \ldots,\ f_n(z)] = 0.$$

Then the equation (2) holds throughout A.

This follows immediately from the fact that the first member of (2) is analytic throughout A.

XXXI

Laurent Series

SERIES OF NEGATIVE POWERS

1. Consider the series

$$(1) \qquad c_1(z-a)^{-1} + \ldots + c_n(z-a)^{-n} + \ldots$$

where the c's and a are constants.

Let us suppose that the series of positive powers

$$(2) \qquad c_1 u + \ldots + c_n u^n + \ldots$$

has a finite radius of convergence $r > 0$. Evidently (1) will converge absolutely when

$$|z - a| > \frac{1}{r}$$

and will diverge when

$$|z - a| < \frac{1}{r}.$$

For $|z - a| > h$, where $h > \frac{1}{r}$, (1) will converge uniformly.

If the radius of convergence of (2) is infinite, (1) converges absolutely and uniformly outside any circle with center at a.

If (2) diverges except for $u = 0$, (1) diverges everywhere. (Note that $(z-a)^{-1}$ cannot be zero.)

STATEMENT OF THE EXPANSION THEOREM OF LAURENT

2. Let $f(z)$ be analytic in the annular open region bounded by two concentric circles C_1 and C_2 whose common center is a point a. We assume C_1 to be interior to C_2. An open region of the type just described will be called a *ring*.

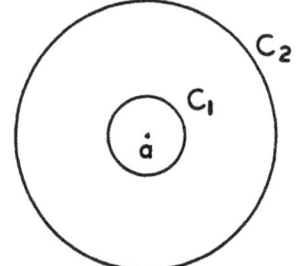

We shall show that $f(z)$ admits in the ring a representation

$$f(z) = c_0 + c_1(z-a) + c_2(z-a)^2 + \ldots$$
$$+ c_{-1}(z-a)^{-1} + c_{-2}(z-a)^{-2} + \ldots$$

The series $\sum_{n=0}^{\infty} c_n(z-a)^n$ converges within C_2, whereas the series of negative power $\sum_{n=-1}^{-\infty} c_n(z-a)^n$ converges outside of C_1.

The above representation of $f(z)$ is unique. The c's are determined uniquely by the formula

$$c_n = \frac{1}{2\pi i} \int_\Gamma \frac{f(z)}{(z-a)^{n+1}} dz$$

where Γ is any circle, with a as center, lying within the ring. The integration is in the positive sense.

3. We shall need the following result.

If n is an integer greater than unity, the integral of $1/z^n$ around any circle with the origin as center is zero.

Proof: Let Γ be any circle with center at the origin. Let Γ^1 be another such circle, of larger radius than Γ. Now $1/z^n$ is analytic in the open region consisting of the entire plane with the origin deleted. The open region just mentioned contains the circles Γ, Γ^1 and the ring bounded by them.

Hence, by the Cauchy Integral Theorem,

$$\int_{\Gamma^1} \frac{dz}{z^n} - \int_{\Gamma} \frac{dz}{z^n} = 0,$$

so that

$$\int_{\Gamma} \frac{dz}{z^n} = \int_{\Gamma^1} \frac{dz}{z^n}.$$

Let r be the radius of Γ^1. Then

$$\left| \int_{\Gamma^1} \frac{dz}{z^n} \right| \leq \frac{1}{r^n} 2\pi r = \frac{2\pi}{r^{n-1}}.$$

Hence, for r arbitrarily large,

$$\left| \int_{\Gamma} \frac{dz}{z^n} \right| \leq \frac{2\pi}{r^{n-1}}.$$

As $n > 1$, we see that $\int_{\Gamma} \frac{dz}{z^n} = 0$.

Suppose now that we have an expansion for $f(z)$

$$(3) \qquad f(z) = \sum_{n=0}^{\infty} c_n (z-a)^n + \sum_{n=-1}^{-\infty} c_n (z-a)^n$$

which is valid in the ring between C_1 and C_2.

Let Γ be any circle, with a as center, lying within the ring. As the two infinite series in (3) converge throughout the ring, each of them will be uniformly convergent along Γ.

Let p be any integer, positive or negative. Then, by (3),

$$(4) \qquad \frac{f(z)}{(z-a)^{p+1}} = \sum_{n=0}^{\infty} c_n (z-a)^{n-p-1} + \sum_{n=-1}^{-\infty} c_n (z-a)^{n-p-1}$$

and the two series in the second member converge uniformly along Γ.

Now

$$(5) \quad \int_{\Gamma} \frac{f(z)\, dz}{(z-a)^{p+1}} = \int_{\Gamma} \sum_{n=0}^{\infty} c_n (z-a)^{n-p-1}\, dz + \int_{\Gamma} \sum_{n=-1}^{-\infty} c_n (z-a)^{n-p-1}\, dz.$$

Because of the uniform convergence mentioned above, we can integrate the series in (5) term by term along Γ, so that

$$\int_{\Gamma} \frac{f(z)\, dz}{(z-a)^{p+1}} = \sum_{n=0}^{\infty} c_n \int_{\Gamma} (z-a)^{n-p-1}\, dz + \sum_{n=-1}^{-\infty} c_n \int_{\Gamma} (z-a)^{n-p-1}\, dz.$$

Now, if $n-p-1 \geq 0$, $(z-a)^{n-p-1}$ is analytic everywhere, so that

$$\int_{\Gamma} (z-a)^{n-p-1}\, dz = 0.$$

If $n-p-1 < -1$, then, as was seen above,

$$\int_{\Gamma} (z-a)^{n-p-1}\, dz = 0.$$

When n−p−1 = −1, that is, when n = p, we have

$$\int_T (z-a)^{n-p-1}\,dz = \int_T \frac{dz}{z-a} = 2\pi i.$$

Thus,

$$\int_T \frac{f(z)\,dz}{(z-a)^{p+1}} = 2\pi i\,c_p$$

or

$$c_p = \frac{1}{2\pi i} \int_T \frac{f(z)\,dz}{(z-a)^{p+1}},$$

which is the result stated in §2. This shows that the Laurent expansion, if it exists, is unique.

DERIVATION OF EXPANSION

4. Consider any ring R bounded by circles D_1 and D_2, each with center at <u>a</u> and lying in the ring bounded by C_1 and C_2. It is understood that D_1 lies within D_2.

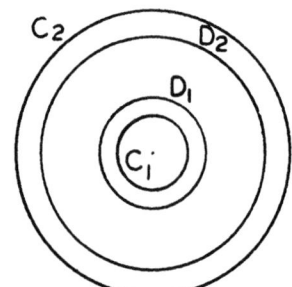

Then $f(z)$ is analytic in an open region containing D_1, D_2 and the ring R bounded by them. Hence, for z in R,

$$f(z) = \frac{1}{2\pi i} \int_{D_2} \frac{f(\zeta)\,d\zeta}{\zeta - z} - \frac{1}{2\pi i} \int_{D_1} \frac{f(\zeta)\,d\zeta}{\zeta - z},$$

the integrations being performed in the positive sense.

Now the expression

$$\frac{1}{2\pi i} \int_{D_2} \frac{f(\zeta)\,d\zeta}{\zeta - z},$$

since $f(\zeta)$ is continuous along D_2, defines an analytic function within D_2. This analytic function has an expansion

(6) $$c_0 + c_1(z-a) + \ldots + c_n(z-a)^n + \ldots,$$

where

$$c_0 = \frac{1}{2\pi i} \int_{D_2} \frac{f(\zeta)\,d\zeta}{\zeta - a}, \qquad c_n = \frac{1}{2\pi i} \int_{D_2} \frac{f(\zeta)\,d\zeta}{(\zeta - a)^{n+1}}, \qquad n \geq 1.$$

This because c_0 in (6) is the value of the analytic function at <u>a</u> and c_n, for $n \geq 1$ is the nth derivative at <u>a</u> divided by n!

We examine now the expression

$$-\frac{1}{2\pi i} \int_{D_1} \frac{f(\zeta)\,d\zeta}{\zeta - z}$$

which equals

$$\frac{1}{2\pi i} \int_{D_1} \frac{f(\zeta)\,d\zeta}{z - \zeta}.$$

We have

$$\frac{1}{z - \zeta} = \frac{1}{(z-a)-(\zeta-a)} = \frac{1}{z-a} + \frac{\zeta-a}{(z-a)^2} + \ldots$$

$$+ \frac{(\zeta-a)^{n-1}}{(z-a)^n} + \frac{(\zeta-a)^n}{(z-\zeta)(z-a)^n}.$$

Hence

$$\frac{1}{2\pi i} \int_{D_1} \frac{f(\zeta)\,d\zeta}{z-\zeta} = \frac{(z-a)^{-1}}{2\pi i} \int_{D_1} f(\zeta)\,d\zeta + \ldots + \frac{(z-a)^{-n}}{2\pi i} \int_{D_1} f(\zeta)(\zeta-a)^{n-1}\,d\zeta$$

$$+ \frac{(z-a)^{-n}}{2\pi i} \int_{D_1} \frac{f(\zeta)(\zeta-a)^n}{z-\zeta}\,d\zeta.$$

Let
$$c_{-1} = \frac{1}{2\pi i} \oint_{D_1} f(\zeta)\, d\zeta, \ldots, c_{-n} = \frac{1}{2\pi i} \oint_{D_1} f(\zeta)\, (\zeta - a)^{n-1}\, d\zeta.$$

Then
$$\frac{1}{2\pi i} \oint_{D_1} \frac{f(\zeta)\, d\zeta}{z - \zeta} = c_{-1}(z-a)^{-1} + \ldots + c_{-n}(z-a)^{-n} + \frac{(z-a)^{-n}}{2\pi i} \oint_{D_1} \frac{f(\zeta)\, (\zeta-a)^n\, d\zeta}{z - \zeta}$$

Hence
$$\left| \frac{1}{2\pi i} \oint_{D_1} \frac{f(\zeta)\, d\zeta}{z - \zeta} - c_{-1}(z-a)^{-1} - \ldots - c_{-n}(z-a)^{-n} \right| = \frac{|z-a|^{-n}}{2\pi} \left| \oint_{D_1} \frac{f(\zeta)\, (\zeta-z)^n\, d\zeta}{z - \zeta} \right|.$$

Let $M > 0$ be such that $|f(\zeta)| < M$ along D_1. Let $\delta > 0$ be such that $|z - \zeta| > \delta$ for ζ on D_1. Note that we are dealing with a fixed value of z, so that such a δ exists. Let r be the radius of D_1. Then
$$\frac{|z-a|^{-n}}{2\pi} \left| \int_{D_1} \frac{f(\zeta)\, (\zeta-a)^n}{z - \zeta}\, d\zeta \right| \leq \frac{|z-a|^{-n}}{2\pi} \frac{Mr^n}{\delta} 2\pi r = \frac{Mr}{\delta} \left(\frac{r}{|z-a|} \right)^n.$$

As z is outside D_1, we have $|z - a| > r$ so that
$$\frac{r}{|z - a|} < 1.$$

It follows immediately that, for z outside D_1, the infinite series
$$c_{-1}(z-a)^{-1} + \ldots + c_{-n}(z-a)^{-n} + \ldots$$

converges to
$$\frac{1}{2\pi i} \oint_{D_1} \frac{f(\zeta)\, d\zeta}{z - \zeta}.$$

Thus, for z in R,
$$f(z) = c_0 + c_1(z-a) + \ldots + c_n(z-a)^n + \ldots$$
$$+ c_{-1}(z-a)^{-1} + \ldots + c_{-n}(z-a)^{-n} + \ldots$$

As R is in any ring at all lying within the original ring, we see by §3 that the above expression for $f(z)$ is valid throughout the original ring.

XXXII

Singularities of Analytic Functions

FUNCTIONS BOUNDED IN THE NEIGHBORHOOD OF A POINT

1. Consider an open region A consisting of the interior of a circle with the center of the circle removed.

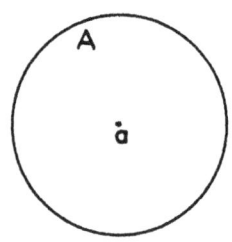

Suppose that $f(z)$ is analytic throughout A and bounded throughout A.

We are going to prove that, if $f(z)$ is defined properly at $\underline{a}$, it will be analytic at $\underline{a}$ as well as in A. That is, by attributing an appropriate value to $f(z)$, for $z = a$, we will have a function analytic throughout the interior of the circle.

Proof: Since $f(z)$ is analytic in every ring obtained by removing from A the centerless interior of any circle with center at $\underline{a}$, $f(z)$ must have a Laurent expansion in powers of $z - a$, valid throughout A.

Suppose that in A,

$$f(z) = c_0 + c_1(z - a) + c_2(z - a)^2 + \ldots$$
$$+ c_{-1}(z - a)^{-1} + c_{-2}(z - a)^{-2} + \ldots.$$

We have

$$c_{-n} = \frac{1}{2\pi i} \int_\Gamma \frac{f(z)\,dz}{(z-a)^{-n+1}} = \frac{1}{2\pi i} \int_\Gamma f(z)\,(z-a)^{n-1}\,dz$$

where Γ is any circle, with center at $\underline{a}$, lying in A.

Let r be the radius of Γ. Let M be an upper bound for $|f(z)|$ in A.
Then

$$|c_{-n}| \leq \frac{1}{2\pi} Mr^{n-1} \cdot 2\pi r = Mr^n.$$

Now, as $n \geq 1$, and as r can be taken arbitrarily small, we see that

$$c_{-n} = 0.$$

Hence

$$f(z) = c_0 + c_1(z - a) + c_2(z - a)^2 + \ldots$$

throughout A. Consequently, if we define $f(z)$ as c_0 at $\underline{a}$, $f(z)$ will be analytic throughout the interior of the circle which forms the outer boundary of A.

ISOLATED SINGULARITIES

2. Suppose that, in A as described above, $f(z)$ is analytic, with a Laurent expansion about $\underline{a}$ which actually contains negative powers.

We say, in that case, that $f(z)$ has an *isolated singularity* at $\underline{a}$ or that $\underline{a}$ is an *isolated singularity* or an *isolated singular point* of $f(z)$.

If the expansion at $\underline{a}$ contains only a finite number of negative powers, we call $\underline{a}$ a *pole* of $f(z)$. If $f(z)$ has a pole at $\underline{a}$ and if the expansion at $\underline{a}$ is

$$c_{-n}(z - a)^{-n} + c_{-n+1}(z - a)^{-n+1} +\ldots+ c_0 + c_1(z - a) +\ldots,$$

with c_{-n} not zero, we call $\underline{a}$ a *pole of order* n *of* $f(z)$.

If an infinite number of negative powers are present in the expansion, we call $\underline{a}$ an *isolated essential singularity* of $f(z)$.

POLES

3. Suppose that $f(z)$ has a pole of order n at $\underline{a}$, so that for a neighborhood of $\underline{a}$,

$$f(z) = c_{-n}(z - a)^{-n} +\ldots+ c_0 + c_1(z - a) +\ldots$$

with $c_{-n} \neq 0$.

We are going to prove that, *for every* $G > 0$, *large at pleasure, a* $\delta > 0$ *can be found such that if* $|z - a| < \delta$, *then* $|f(z)| > G$.

That is, for values of z close to $\underline{a}$, the modulus of $f(z)$ is very large.

Proof: For $z \neq a$ and close to $\underline{a}$, we have

$$(1) \qquad f(z) = \frac{c_{-n} + c_{-n+1}(z - a) +\ldots}{(z - a)^n}.$$

The series of non-negative powers in the numerator of the second member of (1) represents a function analytic in some circle with $\underline{a}$ as center. This function is continuous at $\underline{a}$ and its value at $\underline{a}$ is c_{-n}. Thus, for $|z - a|$ small,

$$|c_{-n} + c_{-n+1}(z - a) +\ldots| > \frac{|c_{-n}|}{2},$$

so that for $|z - a|$ small,

$$|f(z)| > \frac{|c_{-n}|}{2} \frac{1}{|z - a|^n}.$$

As $n > 0$, $|f(z)|$ is very large when $|z - a|$ is very small.

Q.E.D.

THE NUMBER INFINITY

4. We are going to adjoin to our complex number system a symbol ∞, dogmatically setting down the following rules for operation with the symbol.

(A) $\frac{1}{\infty} = 0$.

(B) If $a \neq 0$, $\frac{a}{0} = \infty$.

(C) If $a \neq 0$, $a \infty = \infty a = \infty$.

(D) If $a \neq \infty$, $\frac{\infty}{a} = \infty$.

(E) If $a \neq \infty$, $a + \infty = \infty + a = \infty$.

(F) If $a \neq \infty$, $\infty - a = \infty$ and $a - \infty = \infty$.

By a *neighborhood of* ∞, we shall mean the exterior of any circle.

The statement "$f(z)$ approaches ∞ as z approaches $\underline{a}$" will mean that $f(z)$ is defined for a neighborhood of $\underline{a}$, except perhaps at $\underline{a}$, that $f(z)$ is finite when z is close to $\underline{a}$ and distinct from $\underline{a}$, and that, for every $G > 0$, a neighborhood of $\underline{a}$ can be found in which $|f(z)| > G$ for $z \neq \underline{a}$.

The modulus of ∞ is defined as ∞. We speak of the modulus of ∞ (which equals ∞), as being "greater than" any real number. This is merely a turn of language which is convenient in certain situations.

If $f(z)$ has a pole at $\underline{a}$, we attribute to $f(z)$ at $\underline{a}$ the value ∞.

RECIPROCALS OF FUNCTIONS

5. Suppose that $f(z)$, not identically zero, has a zero at $\underline{a}$. Let $f(z)$ have at $\underline{a}$ the expansion

$$f(z) = c_n(z - a)^n + c_{n+1}(z - a)^{n+1} + \ldots$$

where $c_n \neq 0$.

Then

$$f(z) = c_n (z - a)^n g(z)$$

where

$$g(z) = 1 + \frac{c_{n+1}}{c_n} (z - a) + \ldots$$

For z close to $\underline{a}$, but distinct from $\underline{a}$,

$$\frac{1}{f(z)} = \frac{1}{c_n(z - a)^n} \frac{1}{g(z)}.$$

As $g(a) = 1 \neq 0$, $\frac{1}{g(z)}$ is analytic and equal to 1 at $\underline{a}$, and we have, for $|z - a|$ small,

$$\frac{1}{g(z)} = 1 + d_1(z - a) + d_2(z - a)^2 + \ldots.$$

Hence

$$\frac{1}{f(z)} = \frac{1}{c_n(z - a)^n} + \frac{d_1}{c_n(z - a)^{n-1}} + \ldots,$$

and $\frac{1}{f(z)}$ has a pole of order n at $\underline{a}$.

ISOLATED ESSENTIAL SINGULARITIES

6. Let $f(z)$ have an isolated essential singularity at a point $\underline{a}$. We shall prove the following theorem of Weierstrass:

Given any complex number h, any $\varepsilon > 0$, and any $\delta > 0$, we can find a z such that $|z-a| < \delta$ and that $|f(z) - h| < \varepsilon$.

Proof: Suppose that there exist an h, an $\varepsilon > 0$ and a $\delta > 0$, such that, for $|z - a| < \delta$, we have $|f(z) - h| \geq \varepsilon$.

Consider the function

$$g(z) = \frac{1}{f(z) - h}.$$

It is analytic and bounded for $|z - a| < \delta$, because $|f(z) - h| \geq \varepsilon$ for $|z - a| < \delta$.

Then, if $g(z)$ is properly defined at $\underline{a}$, $g(z)$ will be analytic at $\underline{a}$.

If

$$g(a) \neq 0,$$

then

$$f(z) - h = \frac{1}{g(z)}$$

will be analytic at $\underline{a}$, so that $f(z)$ will be analytic at $\underline{a}$. This is absurd.

If $g(a) = 0$, then $f(z) - h$ has a pole at $\underline{a}$. This means that $f(z)$ has a pole at $\underline{a}$, which contradicts the hypothesis.

Q.E.D.

7. Let $f(z)$ be analytic outside some circle C with center at the origin.

Then $f(z)$ is analytic in the ring contained between the circle C and a circle of arbitrarily large radius.

Thus $f(z)$ has throughout the exterior of the circle C a representation

$$f(z) = c_0 + c_1 z + c_2 z^2 + \ldots$$
$$+ \frac{c_{-1}}{z} + \frac{c_{-2}}{z^2} + \ldots$$

This expansion is called the *Laurent expansion of $f(z)$ about ∞*.

If this expansion contains no positive powers of z, that is, if every c_n with $n > 0$ is zero, $f(z)$ is said to be *analytic at ∞*.

If positive powers are actually present, $f(z)$ is said to have an isolated singularity at ∞. If only a finite number of positive powers are present, and if the highest such power is z^n, $f(z)$ is said to have a *pole of order n at ∞*.

If an infinite number of positive powers are present, $f(z)$ is said to have an *essential singularity at ∞*.

The behavior of $f(z)$ at ∞ thus has the same description as the behavior of $f(\frac{1}{u})$ for $u = 0$.

If $f(z)$ is analytic at ∞, $f(z)$ tends toward a finite limit as z approaches ∞. If $f(z)$ has a pole at ∞, $f(z)$ approaches ∞ as z approaches ∞. If $f(z)$ has an essential singularity at ∞, $f(z)$ comes arbitrarily close to every value in every neighborhood of ∞.

XXXIII

Products and Quotients of Analytic Functions

MULTIPLICITY OF A VALUE

1. Let a non-constant function $f(z)$ be analytic at a finite point $\underline{a}$. Let $f(z)$ have, for the neighborhood of $\underline{a}$, an expansion

$$c_0 + c_p(x - a)^p + c_{p+1}(x - a)^{p+1} + \ldots$$

where $c_p \neq 0$. We assume, that is, that $f(a) = c_0$ and that if $p > 1$,

$$f'(a) = 0, \; f''(a) = 0, \; \ldots, \; f^{(p-1)}(a) = 0; \; f^{(p)}(a) \neq 0.$$

If $p = 1$, the understanding is that $f'(a) \neq 0$.

The function $f(z)$ is then said to assume the value c_0 *p times* at $\underline{a}$.

If $c_0 = 0$, $\underline{a}$ is called a *zero of order p* of $f(z)$.

If a function has a pole of order p at a point $\underline{a}$, the function is said to assume the value ∞ *p times* at $\underline{a}$.

Suppose that $f(z)$ is analytic at ∞. Let it have at ∞ an expansion

$$c_0 + \frac{c_p}{z^p} + \frac{c_{p+1}}{z^{p+1}} + \ldots$$

with $c_p \neq 0$. Then $f(z)$ is said to assume the value c_0 *p times* at ∞. If $c_0 = 0$, $f(z)$ is said to have a *zero of order p* at ∞.

PRODUCT OF TWO FUNCTIONS

2. By a *vicinity* of a finite point $\underline{a}$, we shall mean the open region obtained by removing $\underline{a}$ from the interior of a circle with center at $\underline{a}$.

Let $f(z)$ and $g(z)$, neither identically zero, be analytic in some vicinity of $\underline{a}$, and let the Laurent expansions of $f(z)$ and $g(z)$ about $\underline{a}$ contain at most a finite number of negative powers of $z - a$, so that $f(z)$ and $g(z)$ are either analytic at $\underline{a}$ or have poles at $\underline{a}$. Let the expansions be

$$f(z) = c_0(z - a)^m + c_1(z - a)^{m+1} + \ldots$$
$$g(z) = d_0(z - a)^n + d_1(z - a)^{n+1} + \ldots$$

where m and n are any integers at all and where $c_0 \neq 0$, $d_0 \neq 0$.

Then, for some vicinity of $\underline{a}$,

$$f(z) = (z - a)^m \varphi(z), \qquad [\varphi(z) = c_0 + c_1(z - a) + \ldots],$$
$$g(z) = (z - a)^n \psi(z), \qquad [\psi(z) = d_0 + d_1(z - a) + \ldots].$$

Thus

$$f(z)\, g(z) = (z - a)^{m+n} \varphi(z)\, \psi(z).$$

Now $\varphi(z)\, \psi(z)$ is analytic at $\underline{a}$ and equal at $\underline{a}$ to $c_0 d_0$ which is not zero.

Thus
$$f(z)\,g(z) = (z-a)^{m+n}[c_0 d_0 + e_1(z-a) + \ldots]$$
in the vicinity.

Thus, if $m+n = 0$, $f(z)\,g(z)$, if defined at $\underline{a}$ as $c_0 d_0$, will be analytic and distinct from 0 at $\underline{a}$.

If $m+n > f(z)\,g(z)$, if defined as 0 at $\underline{a}$, will have a zero of order $m+n$ at $\underline{a}$.

If $m+n < 0$, $f(z)\,g(z)$ will have a pole of order $-(m+n)$ at $\underline{a}$.

Suppose that $f(z)$ and $g(z)$, neither identically zero, are analytic for a neighborhood of ∞ and have no essential singularity at ∞. Let their expansions at ∞ be
$$f(z) = c_0 z^m + c_1 z^{m-1} + \ldots$$
$$g(z) = d_0 z^n + d_1 z^{n-1} + \ldots$$
with $c_0 \neq 0$, $d_0 \neq 0$.

One proves easily that $f(z)\,g(z)$ has at ∞ a development
$$f(z)\,g(z) = z^{m+n}\left(c_0 d_0 + \frac{e_1}{z} + \ldots\right)$$
and draws conclusions as above.

Of course, if either $f(z)$ or $g(z)$ is identically zero, then $f(z)\,g(z)$ is identically zero.

QUOTIENT OF TWO FUNCTIONS

3. Let $f(z)$ and $g(z)$, both analytic for a vicinity of a finite point $\underline{a}$, and neither identically zero, habe at $\underline{a}$ expansions
$$f(z) = c_0(z-a)^m + c_1(z-a)^{m+1} + \ldots$$
$$g(z) = d_0(z-a)^n + d_1(z-a)^{n+1} + \ldots$$
with m and n positive, negative or zero. We understand that $c_0 \neq 0$, $d_0 \neq 0$.

We consider a vicinity of $\underline{a}$, contained in the given one, in which $g(z)$ has no zero. Such a neighborhood exists, because if $g(z)$ has a pole at $\underline{a}$, $|g(z)|$ is large for z close to $\underline{a}$, and if $g(z)$ has no pole at $\underline{a}$, $g(z)$ is analytic at $\underline{a}$ if properly defined at $\underline{a}$, so that $\underline{a}$ cannot be a limit point of zeros of $g(z)$.

In the new vicinity,
$$\frac{f(z)}{g(z)}$$
is analytic.

We have
$$g(z) = (z-a)^n[d_0 + d_1(z-a) + \ldots],$$
so that in the new vicinity,
$$\frac{1}{g(z)} = \frac{1}{(z-a)^n}\,\frac{1}{d_0 + d_1(z-a) + \ldots}.$$

Now the series $d_0 + d_1(z-a) + \ldots$ is analytic and distinct from 0 at $\underline{a}$, as well as in the new vicinity. Thus, in the new neighborhood,
$$\frac{1}{g(z)} = \frac{1}{(z-a)^n}\left(\frac{1}{d_0} + e_1(z-a) + \ldots\right).$$

We have thus

$$\frac{1}{g(z)} = \frac{1}{d_0}(z-a)^{-n} + e_1(z-a)^{-n+1} + \ldots$$

so that

$$\frac{f(z)}{g(z)} = \frac{c_0}{d_0}(z-a)^{m-n} + h_1(z-a)^{m-n+1} + \ldots .$$

Thus, if $m - n = 0$, $\frac{f(z)}{g(z)}$, when properly defined at a, is analytic and distinct from zero at a. If $m - n > 0$, $\frac{f(z)}{g(z)}$, if defined as 0 at a, will have a zero of order $m - n$ at a. If $m - n < 0$, $\frac{f(z)}{g(z)}$ will have a pole of order $n - m$ at a.

Similar results hold for functions $f(z)$ and $g(z)$ which are analytic in a neighborhood of ∞ and do not have an essential singularity at ∞.

PRINCIPAL PART OF A LAURENT DEVELOPMENT

4. Let $f(z)$ have an isolated singularity at the finite point a, so that for a vicinity of a,

$$f(z) = c_0 + c_1(z-a) + \ldots + c_n(z-a)^n + \ldots$$

$$+ c_{-1}(z-a)^{-1} + \ldots + c_{-n}(z-a)^{-n} + \ldots .$$

The series of negative powers

$$c_{-1}(z-a)^{-1} + \ldots + c_{-n}(z-a)^{-n} + \ldots$$

is called the *principal part* of the Laurent development at a.

Let $f(z)$ have an isolated singularity at ∞, with a Laurent development

$$c_0 + c_1 z + \ldots + c_n z^n + \ldots$$

$$+ c_{-1} z^{-1} + \ldots + c_{-n} z^{-n} + \ldots .$$

We call the series of *positive powers*

$$c_1 z + \ldots + c_n z^n + \ldots$$

the *principal part* of the development.

XXXIV

Rational Functions

1. By a *rational function*, we shall mean a function defined by an expression

$$\frac{a_0 z^m + \ldots + a_m}{b_0 z^n + \ldots + b_n}.$$

Here m and n are non-negative integers. The a_i and b_i are complex numbers. It is understood that not all of the b_i are zero.

At all finite points except the points at which the denominator vanishes, which latter points are at most n in number, the expression defines an analytic function. Where the denominator vanishes, and at ∞, we attribute a meaning to the expression in accordance with the principles developed in the preceding set of notes.

Thus a rational function is analytic except perhaps at a finite number of points at which it has poles.

We prove the theorem:

Theorem: A rational function devoid of poles is a constant.

Proof: Such a function is an integral function. Having no pole at ∞, it is bounded in some neighborhood of ∞. Certainly it is bounded within any finite circle. Thus, it is bounded all over the finite plane (the totality of finite complex numbers) and hence is a constant.

DECOMPOSITION OF A RATIONAL FUNCTION INTO PARTIAL FRACTIONS

2. Let the rational function $f(z)$ have a pole at ∞, with a principal part

$$A_1 z + A_2 z^2 + \ldots + A_p z^p$$

in its development at ∞.

Then the rational function

$$f(z) - (A_1 z + \ldots + A_p z^p),$$

if properly defined at ∞, will be analytic at ∞, and will have poles in the finite part of the plane only where $f(z)$ does.

Consider now a rational function $f(z)$, analytic at ∞, which has poles at a finite number of finite points

$$h_1, \ldots, h_r,$$

the principal parts of its development being respectively

$$g_1(z) = \frac{A_{11}}{(z - h_1)} + \ldots + \frac{A_{1p_1}}{(z - h_1)^{p_1}}$$

$$\cdot \quad \cdot \quad \cdot \quad \cdot \quad \cdot \quad \cdot \quad \cdot$$

$$g_r(z) = \frac{A_{r1}}{(z - h_r)} + \ldots + \frac{A_{rp_r}}{(z - h_r)^{p_r}}.$$

Then each $g_i(z)$ is analytic everywhere, even at ∞, except at h_i.

Hence
$$f(z) - g_1(z) - \ldots - g_r(z)$$
is a rational function devoid of poles, and is thus a constant.

We have thus the following theorem on the decomposition of a rational function into partial fractions:

Every rational function can be represented in the form
$$A_0 + \ldots + A_p z^p + \sum_{i=1}^{r} \left[\frac{A_{i1}}{z - h_i} + \ldots + \frac{A_{i p_i}}{(z - h_i)^{p_i}} \right]$$
where $h_1, \ldots, h_r$ are the finite poles of the function. The representation is unique.

FUNCTIONS WHOSE ONLY SINGULARITIES ARE POLES

3. By the *extended plane*, we shall mean the entire set of complex numbers, including ∞.

Theorem: A function which is analytic in the extended plane, except perhaps for points at which it has poles, is a rational function.

Our first step will be to prove that the function, call it $f(z)$, has at most a finite number of poles.

Inside any circle, $f(z)$ can have only a finite number of poles. Otherwise the poles would have a limit point within or on the circle. Such a limit point could be neither a pole nor a point of analyticity of $f(z)$, for in a sufficiently small vicinity of a point of these types, $f(z)$ is analytic. Again, there must exist a neighborhood of ∞ in which $f(z)$ has no pole, for if $f(z)$ is analytic at ∞ or has a pole at ∞, $f(z)$ will be analytic in a neighborhood of ∞.

Thus $f(z)$ can have at most a finite number of poles. We know from the preceding section that $f(z)$ equals the sum of the principal parts of the Laurent developments at all of its poles, plus a constant. Thus $f(z)$ is a rational function, since each principal part is rational.

FREQUENCY OF ATTAINMENT OF VALUES

4. If, in a certain domain, a function assumes a certain value h only at a finite number of points, say at the points
$$r_1, \ldots, r_q$$
and if h is assumed p_i times at r_i, then $f(z)$ is said to assume the value h
$$p_1 + \ldots + p_q$$
times in the domain. If h is zero, the function is said to have $p_1 + \ldots + p_q$ zeros in the domain. If $h = \infty$, the function is said to have $p_1 + \ldots + p_q$ poles in the domain.

If $f(z)$ assumes the value h n times in a domain, where h is finite, then $f(z) - h$ has n zeros in the domain.

POLYNOMIALS

5. By a polynomial, we mean a function defined by an expression
$$(1) \qquad a_0 z^m + a_1 z^{m-1} + \ldots + a_m.$$

If the polynomial is not identically zero, we may write its expression so that $a_0 \neq 0$. In that case, we call m the *degree* of the polynomial. We do not attribute a degree to the polynomial zero.

If the polynomial is of degree $m > 0$, we can, by the fundamental theorem of algebra, write it in the form

$$a_0(z - r_1)^{p_1} \ldots (z - r_q)^{p_q}$$

where $r_1, \ldots, r_q$ are distinct from one another and where

$$p_1 + \ldots + p_q = m.$$

Then the polynomial has p_i zeros at r_i.

Hence a polynomial of degree $m > 0$ has m zeros.

(Note that the polynomial, if $m > 0$, has a pole at ∞, so that $r_1, \ldots, r_q$ are the only zeros.)

The theorem just stated evidently holds also for $m = 0$.

DEGREE OF A RATIONAL FUNCTION

6. Consider a rational function

$$f(z) = \frac{a_0 z^m + \ldots + a_m}{b_0 z^n + \ldots + b_n}$$

which is not identically zero, that is, one for which $a_0, \ldots, a_m$ do not all vanish.

We suppose any linear factors common to the numerator and denominator to be removed, so that the two polynomials have no zero in common.

Thus, at any point, necessarily finite, at which the numerator has a zero of a certain order, $f(z)$ has a zero of the same order. At a point at which the denominator has a zero of a certain order, $f(z)$ has a pole of that order.

Thus, in the finite part of the plane, $f(z)$ has just m zeros and just n poles.

If $m = n$, ∞ is neither a zero nor a pole of $f(z)$ and $f(z)$ *has as many zeros as poles*.

If $m > n$, $f(z)$ has $m - n$ poles at ∞. Thus $f(z)$ has $(m - n) + n = m$ poles in all. Again, $f(z)$ has as many zeros as poles.

Finally, if $m < n$, $f(z)$ has $n-m$ zeros at ∞. Thus $f(z)$ has $(n - m) + m = n$ zeros in all, so that $f(z)$ has as many zeros as poles.

We call the greater of the two numbers m and n the *degree* of $f(z)$.

Thus, *a rational function distinct from zero has as many zeros, and as many poles, as there are units in its degree.*

Evidently a rational function of degree zero is a constant.

Let $f(z)$ be a rational function of positive degree. Let h be any finite number. Then $f(z) - h$ has as many poles as $f(z)$ does, because the subtraction of h does not affect the principal parts of the developments at the poles.

Hence, $f(z) - h$ has a number of zeros equal to the degree of $f(z)$.

We formulate this result as follows: *A rational function of degree $p \geq 1$ assumes every complex value exactly p times in the extended plane.*

We do not attribute a degree to the rational function 0.

XXXV

The Functions e^z, sin z, cos z

THE DEFINITIONS

1. We introduce three integral functions which we shall define by means of Taylor series with infinite radii of convergence. The functions are

$$e^z = 1 + z + \frac{z^2}{2!} + \ldots + \frac{z^n}{n!} + \ldots$$

$$\sin z = z - \frac{z^3}{3!} + \frac{z^5}{5!} - \ldots + (-1)^n \frac{z^{2n+1}}{(2n+1)!} + \ldots$$

$$\cos z = 1 - \frac{z^2}{2!} + \frac{z^4}{4!} - \ldots + (-1)^n \frac{z^{2n}}{(2n)!} + \ldots$$

We regard the initial term of each series as a *zero* th term.

That the radius of convergence is infinite in each case is easily shown by the Cauchy-Hadamard theorem.

We notice that when z is real, each of the three functions is real.

FIRST RELATIONS

2. Simple calculations show that

$$e^{iz} = \cos z + i \sin z, \quad e^{-iz} = \cos z - i \sin z.$$

It follows that

$$\sin z = \frac{e^{iz} - e^{-iz}}{2i}, \quad \cos z = \frac{e^{iz} + e^{-iz}}{2}.$$

We find, differentiating, that

$$\frac{d}{dz} e^z = e^z; \quad \frac{d}{dz} \sin z = \cos z; \quad \frac{d}{dz} \cos z = -\sin z.$$

It is easy to see that

$$\sin(-z) = -\sin z; \quad \cos(-z) = \cos z.$$

THE MULTIPLICATION FORMULA FOR e^z

3. *Lemma: Let $f(z)$ be analytic in an open region* A. *Let $f'(z) = 0$ throughout* A. *Then $f(z)$ is a constant.*

Proof: For some point $\underline{a}$ of A, let $f(z)$ have an expression

$$f(z) = c_0 + c_1(z - a) + c_2(z - a)^2 + \ldots$$

Then, for a neighborhood of $\underline{a}$,

$$f'(z) = c_1 + 2c_2(z - a) + \ldots.$$

As $f'(z)$ is identically zero, we must have

$$c_1 = c_2 = c_3 = \ldots = 0.$$

Thus $f(z) = c_0$ for a neighborhood of $\underline{a}$, and hence throughout A.

We prove now the theorem:

Theorem: For every z and h,

$$e^{z+h} = e^z e^h.$$

Proof: We have, considering h as any constant and z as a variable,

$$\frac{d}{dz} e^{-z} e^{z+h} = e^{z+h} \frac{d}{dz} e^{-z} + e^{-z} \frac{d}{dz} e^{z+h}$$

Now

$$\frac{d}{dz} e^{-z} = -e^{-z}; \quad \frac{d}{dz} e^{z+h} = e^{z+h}.$$

Hence

$$\frac{d}{dz} e^{-z} e^{z+h} = -e^{z+h} e^{-z} + e^{-z} e^{z+h} = 0.$$

Then

$$e^{-z} e^{z+h} = C,$$

with C a constant depending on h. We put $z = 0$. Since $e^0 = 1$, we find

$$e^0 e^{0+h} = e^h = C.$$

Thus

(1) $$e^{-z} e^{z+h} = e^h$$

for every z and every h. Putting $h = 0$, we find

(2) $$e^{-z} e^z = 1.$$

Multiplying both sides of (1) by e^z, and having regard to (2), we find

$$e^{z+h} = e^z e^h.$$

Q.E.D.

THE RELATION $SIN^2 z + COS^2 z = 1$

4. We have

$$\sin^2 z = \left(\frac{e^{iz} - e^{-iz}}{2i}\right)^2; \quad \cos^2 z = \left(\frac{e^{iz} + e^{-iz}}{2}\right)^2.$$

Then, by §3,

$$\sin^2 z + \cos^2 z = \frac{e^{2iz} - 2 + e^{-2iz}}{-4} + \frac{e^{2iz} + 2 + e^{-2iz}}{4}$$

or

$$\sin^2 z + \cos^2 z = 1.$$

THE ADDITION THEOREMS FOR SIN z AND COS z

5. Using the expressions for sin z and cos z in terms of e^{iz} and the multiplication theorem for e^z, we find that

$$\sin(z + h) = \sin z \cos h + \cos z \sin h$$
$$\cos(z + h) = \cos z \cos h - \sin z \sin h.$$

PERIODICITY OF SIN z, COS z, e^z

6. Given an integral function $f(z)$, if an $h \neq 0$ exists such that $f(z + h) = f(z)$, we shall say that $f(z)$ is *periodic* and we shall call h a *period of* $f(z)$.

We shall prove that $\sin z$ and $\cos z$ are periodic with 2π for period.

We have $\cos 0 = 1$. Also, by actual calculation, we find $\cos 2 < 0$. Since $\cos z$ is real and continuous in the interval $(0, 2)$, there is a point inside the interval for which $\cos z = 0$. Now, since the zeros of an analytic function which is not identically zero are isolated, there is, among the positive real values of z for which $\cos z = 0$, one value which is a minimum. Let this zero of $\cos z$ be denoted by $\frac{p}{2}$. It will turn out later that $p = \pi$.

Since $\sin^2 z + \cos^2 z = 1$, we have $\sin \frac{p}{2} = \pm 1$. We shall prove that $\sin \frac{p}{2} = +1$. Inside the interval $(0, \frac{p}{2})$ we have $\cos z > 0$. By the mean value theorem,

$$\sin \frac{p}{2} - \sin 0 = (\cos x_1)\left(\frac{p}{2} - 0\right)$$

where $0 < x_1 < \frac{p}{2}$. As $\sin 0 = 0$, we have

$$\sin \frac{p}{2} = \frac{p}{2} \cos x_1 > 0$$

so that $\sin \frac{p}{2} = 1$.

Using the formulas for $\sin(z + h)$ and $\cos(z + h)$ with $h = \frac{p}{2}$, we find that

$$\sin\left(z + \frac{p}{2}\right) = \cos z; \quad \cos\left(z + \frac{p}{2}\right) = -\sin z.$$

Hence
$$\sin(z + p) = \cos\left(z + \frac{p}{2}\right) = -\sin z.$$
and
$$\cos(z + p) = -\sin\left(z + \frac{p}{2}\right) = -\cos z.$$

Finally,
$$\sin(z + 2p) = -\sin(z + p) = \sin z$$
$$\cos(z + 2p) = -\cos(z + p) = \cos z.$$

Thus $\sin z$ and $\cos z$ are periodic, with $2p$ for period.

It follows easily, from §2, that $2pi$ is a period for e^z.

Q.E.D.

OTHER RELATIONS

7. We have
$$\sin\left(\frac{p}{2} - z\right) = \sin\frac{p}{2}\cos(-z) + \cos\frac{p}{2}\sin(-z) = \cos(-z) = \cos z.$$

Similarly,
$$\cos\left(\frac{p}{2} - z\right) = \sin z, \quad \sin(p - z) = \sin z,$$
$$\cos(p - z) = -\cos z, \quad \sin\left(\frac{3p}{2} - z\right) = -\cos z, \text{ etc.}$$

Also
$$e^{z+pi} = -e^z, \quad e^{z+\frac{p}{2}i} = ie^z.$$

142

GRAPHS

8. From the periodicity of sin z and cos z and from the fact that for real values of z, they admit 1 and -1 as maximum and minimum values (use relation $\sin^2 z + \cos^2 z = 1$), we see that the graphs of these functions for real values of z are of the types exhibited below.

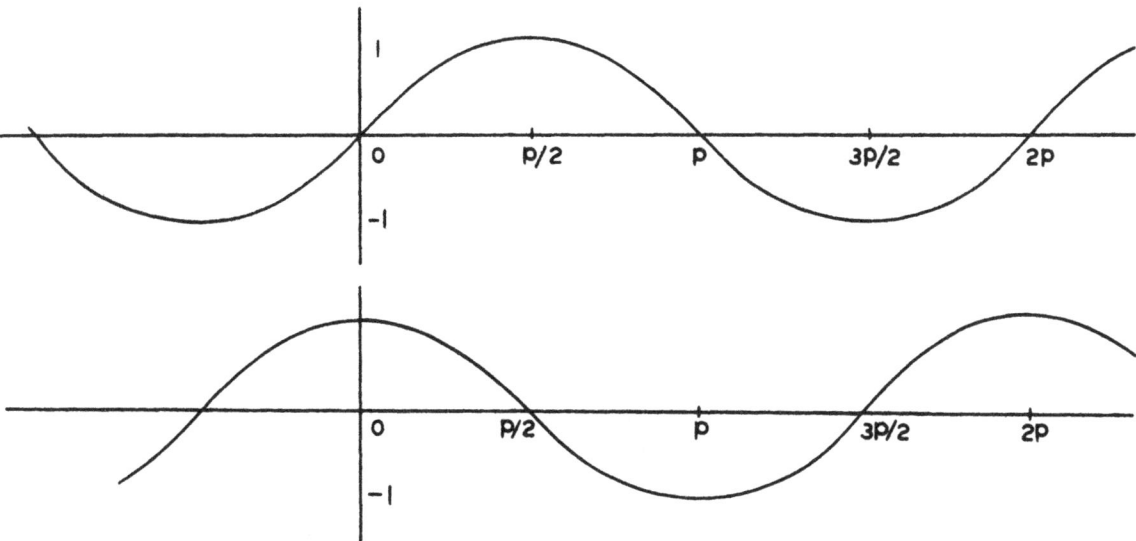

The graph e^z is seen to be

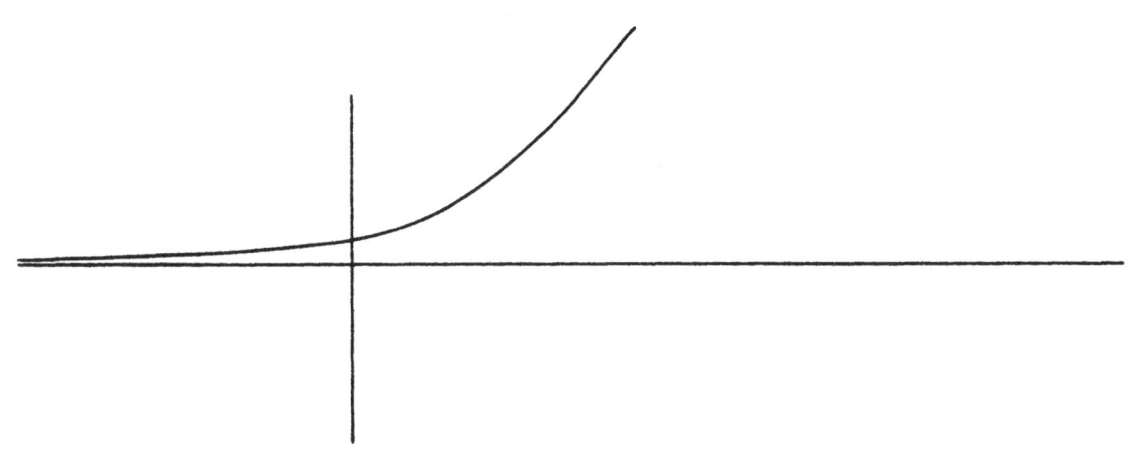

PARAMETRIC REPRESENTATION OF CIRCLE

9. We are going to prove that the curve

$$x = \cos t, \quad y = \sin t \qquad 0 \leq t \leq 2p$$

is a positively sensed circle of radius 1 with center at the origin.

First, because $\cos^2 t + \sin^2 t = 1$, every point of the above curve lies on the circle mentioned.

Consider now any point of the circle. To fix our ideas, let $x \geq 0$, $y \geq 0$. There is a t on the interval $(0, \frac{p}{2})$ such that $\cos t = x$, since $0 \leq x \leq 1$. Now, for $0 \leq t \leq \frac{p}{2}$,

$$\sin t = +\sqrt{1 - \cos^2 t} = +\sqrt{1 - x^2},$$

so that $y = \sin t$. Thus, every point of the circle is on our curve. Finally, it is easy to see that two distinct values of t other than 0 and 2p give distinct points of our curve, so that the curve is a simple closed curve.

Hence, by the theorem on the equivalence or inverse equivalence of two simple closed curves with the same points, our curve is either a positively sensed circle or a negatively sensed circle. Now the sense must be positive, because as t increases from 0 to $\frac{p}{2}$, x decreases and y increases.

A FORMULA FOR THE LENGTH OF A RECTIFIABLE CURVE

10. *Theorem: Let a rectifiable curve C be given by*

$$x = \varphi(t), \quad y = \psi(t) \qquad a \leq t \leq b,$$

where $\varphi(t)$ and $\psi(t)$ have continuous first derivatives on the interval (a, b). Let λ be the length of C. Then

$$\lambda = \int_a^b \sqrt{\left(\frac{d\varphi}{dt}\right)^2 + \left(\frac{d\psi}{dt}\right)^2}\, dt.$$

Proof: We must first show that the integrand is integrable. Towards this end, we prove it continuous. It will suffice to show that the non-negative square root of any non-negative continuous function is continuous.

Let $\alpha(t)$ be continuous on (a, b), with $\alpha(t) \geq 0$. Let $\beta(t) = +\sqrt{\alpha(t)}$. If $\alpha(t_0) = 0$, then, for t close to t_0, $\alpha(t)$ is very small, so that its square root is very small. Hence $\beta(t)$ is continuous at t_0. Suppose $\alpha(t_0) > 0$. Then

$$\beta(t_0 + h) - \beta(t_0) = \sqrt{\alpha(t_0 + h)} - \sqrt{\alpha(t_0)} = \frac{\alpha(t_0 + h) - \alpha(t_0)}{\sqrt{\alpha(t_0 + h)} + \sqrt{\alpha(t_0)}},$$

from which we see again that $\beta(t)$ is close to $\beta(t_0)$ for t close to t_0. This settles the question of integrability.

Now, let (a, b) be divided up by points

$$t_0 = a,\ t_1 > t_0,\ t_2 > t_1,\ \ldots,\ t_n = b.$$

The corresponding inscribed polygon has a length

$$\sqrt{[\varphi(t_1) - \varphi(t_0)]^2 + [\psi(t_1) - \psi(t_0)]^2} + \ldots + \sqrt{[\varphi(t_n) - \varphi(t_{n-1})]^2 + [\psi(t_n) - \psi(t_{n-1})]^2}$$

By the mean value theorem, there are in every interval (t_{i-1}, t_i) points τ_i and σ_i such that

$$\varphi(t_i) - \varphi(t_{i-1}) = \varphi'(\tau_i)(t_i - t_{i-1})$$
$$\psi(t_i) - \psi(t_{i-1}) = \psi'(\sigma_i)(t_i - t_{i-1}).$$

Thus, the length of the polygon is

$$\sum_{i=1}^{n} \sqrt{[\varphi'(\tau_i)]^2 + [\psi'(\sigma_i)]^2}\,(t_i - t_{i-1}).$$

It is easy to show that the function of u and v

$$\sqrt{[\varphi'(u)]^2 + [\psi'(v)]^2}$$

is uniformly continuous with respect to both variables in the closed rectangle $a \leq u \leq b$, $a \leq v \leq b$.

Let an $\varepsilon > 0$ be assigned. Let a $\delta > 0$ be taken in such a way that

$$\left| \sqrt{[\varphi'(u_2)]^2 + [\psi'(v_2)]^2} - \sqrt{[\varphi'(u_1)]^2 + [\psi'(v_1)]^2} \right| < \varepsilon$$

for $|u_2 - u_1| < \delta$ and $|v_2 - v_1| < \delta$. Let δ be also such that

$$\left| \int_a^b \sqrt{[\varphi'(t)]^2 + [\psi'(t)]^2}\, dt - \sum_{i=1}^{n} \sqrt{[\varphi'(t_i)]^2 + [\psi'(t_i)]^2}\, (t_i - t_{i-1}) \right| < \varepsilon$$

if every $t_i - t_{i-1}$ is less than δ. Let every $t_i - t_{i-1}$ be less than δ. Then, since

$$|\sigma_i - t_i| < \delta, \quad |\tau_i - t_i| < \delta,$$

we have

$$\left| \sum_{i=1}^{n} \sqrt{[\varphi'(\tau_i)]^2 + [\psi'(\sigma_i)]^2}\, (t_i - t_{i-1}) - \sum_{i=1}^{n} \sqrt{[\varphi'(t_i)]^2 + [\psi'(t_i)]^2}\, (t_i - t_{i-1}) \right| < \varepsilon(b - a).$$

Hence the length of our polygon differs from

$$\int_a^b \sqrt{[\varphi'(t)]^2 + [\psi'(t)]^2}\, dt$$

by less than $\varepsilon(b - a) + \varepsilon$. This proves the theorem.

PROOF THAT $p = \pi$

11. We know that the length of a circle of radius unity is

$$\int_0^{2p} \sqrt{\cos^2 t + \sin^2 t}\, dt = \int_0^{2p} dt = 2p.$$

Hence

$$2p = 2\pi$$
$$p = \pi$$

Q.E.D.

AMPLITUDES OF A COMPLEX NUMBER

12. Consider any finite complex number $x + yi$, distinct from 0. If r is its modulus, we have

$$x + yi = r\left(\frac{x}{r} + \frac{y}{r}i\right).$$

Now, as

$$\left|\frac{x}{r}\right| \leq 1, \quad \left|\frac{y}{r}\right| \leq 1, \quad \left(\frac{x}{r}\right)^2 + \left(\frac{y}{r}\right)^2 = 1,$$

it is possible to find a real θ such that

$$\cos \theta = \frac{x}{r}, \quad \sin \theta = \frac{y}{r}.$$

We call any such θ an *amplitude* of $x + yi$.

There will be exactly one amplitude θ such that

$$-\pi < \theta \leq \pi.$$

We call it the *principal amplitude* of $x + yi$.

Every amplitude of $x + yi$ differs from the principal amplitude by an integral multiple of 2π.

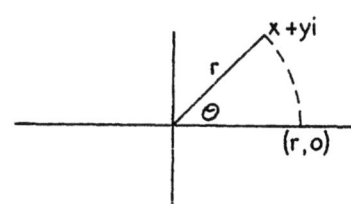

Plotting $x + yi$ in a plane, we can use, for a geometric image of the principal amplitude, the positive or negative angle greater than $-\pi$ and not greater than π which the line joining the origin to $x + yi$ makes with the positive real axis. To justify this convention, let us suppose, to fix our ideas, that $y \geq 0$. Consider the circle of radius r,

$$x = r \cos t, \quad y = r \sin t.$$

The length of that arc of this circle which joins the point $(r, 0)$ to (x, y) is

$$\int_0^\theta r\sqrt{\cos^2 t + \sin^2 t}\, dt = \int_0^\theta r\, dt = r\theta.$$

Thus, our convention is harmonious with our intuitive idea of angle.

XXXVI

Periodic Functions

PERIOD STRIP

1. Consider a periodic integral function f(z) with period h.

Let z be any point of the complex plane. Through z, pass a line which does not go through z + h. Through z + h, pass a line parallel to the first one. Form a point set by adding to the points between the two lines the points on either of the lines (but not both). Such a point set will be called a *period strip* of f(z).

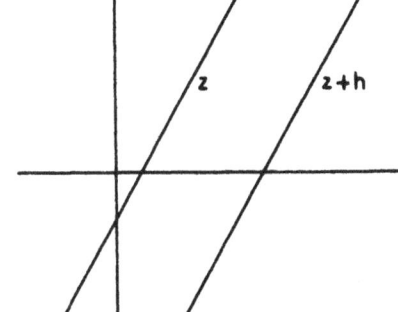

FUNDAMENTAL DOMAIN

2. Two points z_1 and z_2 will be called *congruent modulo* h if $z_2 - z_1$ is an integral multiple of h, that is, if $z_2 - z_1 = mh$ with m an integer. We write

$$z_2 \equiv z_1 \quad (\text{mod. } h).$$

A point set E will be called a *fundamental domain* of f(z) if, for every point z of the plane, there is one and only one point of E which is congruent to z modulo h. This implies, of course, that no two points of E are congruent to each other.

The period strips are fundamental domains.

The behavior of a periodic integral function all over the plane is known as soon as we know the behavior of the function in any fundamental domain.

THE FUNCTION e^z

3. The function e^z has $2\pi i$ as a period. For a period strip, we use the set of points $z = x + yi$ for which $0 \leq y < 2\pi$.

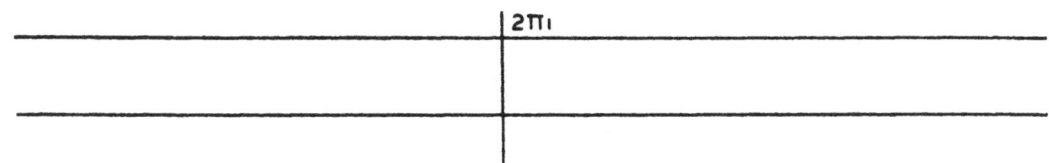

We shall show that e^z assumes every finite value except 0 exactly once in this period strip.

First, because $\frac{1}{e^z} = e^{-z}$ is an integral function, e^z is never zero. (A zero of e^z would be a pole for e^{-z}.)

Consider any finite complex number a + bi, distinct from 0. Let $r = \sqrt{a^2 + b^2}$ and let y, where $0 \leq y < 2\pi$ be such that

$$a + bi = r(\cos y + i \sin y).$$

Since r > 0, there is an x such that $e^x = r$.

Then

$$a + bi = e^x(\cos y + i \sin y) = e^x e^{iy} = e^{x+iy}.$$

Hence, at the point x + iy which lies in our period strip, we have

$$e^{x+iy} = a + bi.$$

To prove that there is only one such point, suppose that

$$e^{x+iy} = a + bi.$$

Then

$$e^x (\cos y + i \sin y) = a + bi.$$

Then e^x, which is positive, must be the modulus of a + bi and y must be one of the amplitudes. There is only one x such that $e^x = |a + bi|$, since e^x is a monotonic function of x. Also, there is only one amplitude of a + bi in the interval $0 \leq y < 2\pi$.

Let $w = e^z$. As z travels along a vertical segment of the period strip, the abscissa of the segment being x, w travels along a circle of radius e^x. This is because, if

$$w = e^{x+iy},$$

with x constant, and y variable on the interval $0 \leq y < 2\pi$, we have, if w = u + iv,

$$u = e^x \cos y, \qquad v = e^x \sin y \qquad 0 \leq y < 2\pi,$$

which equations define a positively sensed circle of radius e^x.

Let the values of w be plotted in a w-plane. That part of the period strip which lies to the left of a vertical segment of abscissa x is carried into the interior of a circle of radius e^x in the w-plane, with the origin, which is the center of the circle, removed.

Thus the function $w = e^z$ maps the period strip upon the w-plane with the origin removed. The mapping is conformal at every point, because the derivative of e^z never vanishes.

We notice that, at points far to the left in the period strip, e^z is very close to zero. (Proof: $|e^z| = e^x$ and e^x is very small if x is large and negative.)

Also, if x is large and positive, $|e^z|$ is large.

We express the fact that $e^z \to 0$ as $x \to -\infty$ by the language: "e^z assumes the value zero at the left end of its period strip." Also, we say "$e^z = \infty$ at the right end of its period strip."

On this basis, we say, "e^z assumes every value exactly once in a period strip."

THE FUNCTIONS SIN z AND COS z

For these functions, we shall use the period strip given by

$$0 \leq x < 2\pi.$$

We have

$$\sin z = \frac{e^{iz} - e^{-iz}}{2i}$$

or

(1) $$\sin z = \frac{e^{2iz} - 1}{2ie^{iz}}.$$

Consider the equation

$$\sin z = a$$

where a is any finite complex number. It becomes, by (1),

$$e^{2iz} - 2aie^{iz} - 1 = 0.$$

148

This quadratic for e^{iz} has two roots, u_1 and $u_2 \neq u_1$ unless $-4a^2 + 4 = 0$, that is, unless $a = \pm 1$. Furthermore, u_1 and u_2 will be distinct from zero.

Suppose that $a \neq \pm 1$.

There is precisely one point of the period strip where $e^{iz} = u_1$ and one where $e^{iz} = u_2$. Thus, the value $a \neq \pm 1$ is assumed by $\sin z$ at two places. We shall see shortly that $\underline{a}$ is assumed only once at each place.

If $a = +1$, we have $e^{iz} = i$, so that $z = \frac{\pi}{2}$. Now, for $z = \frac{\pi}{2}$, $\frac{d}{dz} \sin z = \cos z = 0$ and $\frac{d^2}{dz^2} \sin z = -\sin z \neq 0$. Hence, $\sin z$ assumes the value twice at $\frac{\pi}{2}$. Similarly $\sin z$ assumes the value -1 twice at $\frac{3\pi}{2}$.

To show that if $a \neq \pm 1$, $\underline{a}$ is assumed only once wherever it is assumed, we notice that for $\underline{a}$ to be a multiple value of $\sin z$ at a point z, $\frac{d}{dz} \sin z = \cos z$ must vanish at the point. Now, if $\cos z = 0$, $\sin z = \pm 1$. But $a \neq \pm 1$.

Q.E.D.

When z moves upward in the period strip, e^{iz} tends toward 0. We see from (1) that $\sin z$ tends toward ∞. Also, since

$$\sin z = \frac{1 - e^{-2iz}}{2ie^{-iz}}$$

and since e^{-iz} tends toward 0 as z moves downward in the strip, $\sin z$ tends toward ∞ as z moves downward.

We say on this basis that "$\sin z$ assumes every value exactly twice in its period strip."

Since $\cos z = \sin(z + \frac{\pi}{2})$, a similar result holds for $\cos z$.

XXXVII

Indefinite Integrals, Logarithms

INDEPENDENCE OF AN INTEGRAL OF PATH OF INTEGRATION

1. *Theorem:* Let A be a simply connected open region. Let $f(z)$ be a function analytic throughout A. Let z_1 and z_2 be any two points of A (distinct or coincident). Then

$$\int_{z_1}^{z_2} f(z)\, dz$$

along a rectifiable curve lying in A is independent of the curve.

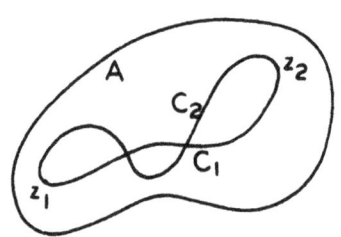

Proof: Let C_1 and C_2 be two rectifiable curve joining z_1 and z_2. Then C_1 and C_2 taken together constitute a *closed* rectifiable curve. Then the integral of $f(z)$ along C_1 from z_1 to z_2, plus the integral from z_2 to z_1 along C_2 equals zero. Now the integral along C_2 from z_2 to z_1 is the negative of the integral along C_2 from z_1 to z_2. Hence

$$\int_{C_1} f(z)\, dz - \int_{C_2} f(z)\, dz = 0$$

where both integrations are made from z_1 to z_2.

THE INDEFINITE INTEGRAL

2. Let A be an open region. Let $f(z)$ be continuous throughout A. Suppose that the integral of $f(z)$ between any two points of A is independent of the path. Let $\underline{a}$ be a fixed point of A. Then

$$\int_a^z f(\zeta)\, d\zeta,$$

the path of integration lying in A, depends on z alone and hence is a function $F(z)$ of z.

We are going to prove that $F(z)$ *is analytic throughout* A *and its derivative is* $f(z)$.

Proof: Let z be any point of A. We keep z fixed during our discussion. Let h be a small complex number, distinct from 0.

In obtaining $F(z + h)$, we may take a path which goes from a to z and then along a straight segment from z to $z + h$. Thus,

$$F(z + h) - F(z) = \int_z^{z+h} f(\zeta)\, d\zeta,$$

the integration being effected along a straight segment. Now $f(z)$ is continuous at z. Taking any $\varepsilon > 0$, let $\delta > 0$ be chosen so that $|f(\zeta) - f(z)| < \varepsilon$ for $|\zeta - z| < \delta$. Then

$$F(z+h) - F(z) = \int_z^{z+h} f(z)\, d\zeta + \int_z^{z+h} [f(\zeta) - f(z)]\, d\zeta = f(z)\,[z + h - z] + \int_z^{z+h} [f(\zeta) - f(z)]\, d\zeta.$$

Thus, if $|h| < \delta$

$$\left|\frac{F(z + h) - F(z)}{h} - f(z)\right| = \left|\frac{1}{h} \int_z^{z+h} [f(\zeta) - f(z)]\, d\zeta\right| \leq \left|\frac{1}{h} \varepsilon h\right| = \varepsilon.$$

This proves that $F(z)$ has a derivative at z and that the derivative is $f(z)$.

The above result leads to *Morera's theorem* which states that:

If f(z) is continuous in an open region and if the integral of f(z) around any closed rectifiable curve in the open region is zero, then f(z) is analytic in the open region.

Proof: First, it is evident that $\int_a^z f(\zeta)\,d\zeta$, where <u>a</u> is a fixed point of the open region and z an arbitrary point, is independent of the path. Let $F(z)$ represent this integral. Then $F(z)$ is analytic, with $f(z)$ for derivative. Thus, $f(z)$ as the derivative of an analytic function is analytic.

LOGARITHM OF AN ANALYTIC FUNCTION

3. Let $f(z)$ be analytic, and nowhere zero, in a simply connected open region A. We shall prove the existence of a function $g(z)$, analytic in A, such that

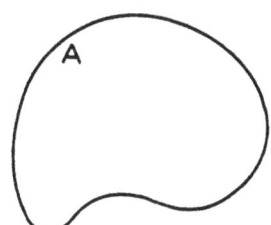

$$f(z) = e^{g(z)}.$$

We shall call $g(z)$ a *logarithm* of $f(z)$ and write $g(z) = \log f(z)$.

Evidently, if we have one such function $g(z)$, we can get an infinite number of other such functions by adding integral multiples of $2\pi i$ to $g(z)$. This is because $e^{2m\pi i} = 1$ if m is in integer. It will be shown that every function $h(z)$, analytic in A, such that $e^{h(z)} = f(z)$, differs from $g(z)$ by an integral multiple of $2\pi i$.

Proof: The function $\dfrac{f'(z)}{f(z)}$ is analytic in A, because $f(z)$ is distinct from 0 throughout A. Then, if <u>a</u> is any fixed point of A,

$$\int_a^z \frac{f'(\zeta)}{f(\zeta)}\,d\zeta$$

is an analytic function of z in A.

Let $\log f(a)$ be any complex number such that $e^{\log f(a)} = f(a)$. Since $f(a) \neq 0$, there are an infinite number of such numbers $\log f(a)$, differing from each other by multiples of $2\pi i$.

Let

$$(1) \qquad g(z) = \log f(a) + \int_a^z \frac{f'(\zeta)}{f(\zeta)}\,d\zeta.$$

We say that $e^{g(z)} = f(z)$.

We have, by (1),

$$\frac{d}{dz} e^{g(z)} = e^{g(z)} \frac{dg(z)}{dz} = e^{g(z)} \frac{f'(z)}{f(z)}.$$

Thus

$$-e^{g(z)} f'(z) + f(z) \frac{d}{dz} e^{g(z)} = 0.$$

As $f(z)$ is nowhere zero in A, we may write

$$\frac{-e^{g(z)} f'(z) + f(z) \dfrac{d}{dz} e^{g(z)}}{[f(z)]^2} = 0.$$

The first member of the last equation is the derivative of

$$\frac{e^{g(z)}}{f(z)}.$$

Hence, $\dfrac{e^{g(z)}}{f(z)}$ is constant throughout A. To determine the constant, we observe that, for $z = a$, $e^{g(z)} = e^{\log f(a)} = f(a)$. Hence, the constant is unity and

$$e^{g(z)} = f(z).$$

If $h(z)$ is an analytic function such that $e^{h(z)} = f(z)$, then evidently, at any particular point z, $h(z)$ differs from $g(z)$ by an integral multiple of $2\pi i$. As $h(z) - g(z)$ is continuous, the integral multiple must be a constant multiple of $2\pi i$, for an integer cannot vary continuously without staying constant.

Q.E.D.

INTEGRAL FUNCTIONS WHICH ARE NOWHERE ZERO

4. The entire finite complex plane is simply connected. Hence, if $f(z)$ is an integral function which is nowhere zero, we have

$$f(z) = e^{g(z)}$$

with $g(z)$ an integral function.

THE DEVELOPMENT OF LOG (1 + z)

5. Consider a circle of radius unity, with center at the origin. Inside this circle, $1 + z$ is nowhere zero, for $1 + z = 0$ only for $z = -1$. Hence, there are functions analytic within the circle whose exponentials equal $1 + z$. We shall call any such function a *branch of log* $(1 + z)$. One of these functions equals 0 for $z = 0$. We call it the *principal branch* of $\log (1 + z)$.

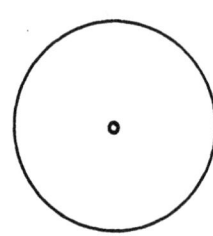

Let us get the Taylor expansion of the principal branch for $z = 0$. We have

$$\frac{d}{dz} \log (1 + z) = \frac{1}{1 + z}, \quad \frac{d^2}{dz^2} \log (1 + z) = -\frac{1}{(1 + z)^2}, \quad \ldots,$$

$$\frac{d^n}{dz^n} \log (1 + z) = \frac{(-1)^{n-1} (n-1)!}{(1 + z)^n}.$$

Then

$$\text{Log } (1 + z) = z - \frac{z^2}{2} + \frac{z^3}{3} - \frac{z^4}{4} + \ldots.$$

XXXVIII

Infinite Products

CONVERGENT INFINITE PRODUCTS

1. By an infinite product of complex numbers, we shall mean a symbol

(1) $$a_1 a_2 \ldots a_n \ldots$$

where each a_n is a complex number.

The infinite product (1) will be said to be *convergent* if both of the following conditions are satisfied:

(α) *At most a finite number of the a's are zero.*

(β) *If the integer m is such that $a_n \neq 0$ for $n \geq m$, then the sequence*

$$a_m, \; a_m a_{m+1}, \; a_m a_{m+1} a_{m+2}, \; \ldots$$

converges and has a limit distinct from zero.

Evidently, if the infinite product is convergent in the above sense, the sequence

$$a_1, \; a_1 a_2, \; a_1 a_2 a_3, \; \ldots$$

converges. The limit of this last sequence is taken as the value of the product (1).

An infinite product which is not convergent is called *divergent*. For instance,

$$1 \cdot 2 \cdot 3 \cdot 4 \ldots$$
$$1 \cdot 0 \cdot 2 \cdot 0 \cdot 3 \cdot 0 \ldots$$
$$\frac{1}{2} \cdot \frac{1}{3} \cdot \frac{1}{4} \cdot \frac{1}{5} \ldots$$

are divergent.

The above definition is so framed as to allow a convergent product to be zero only if one of its factors is zero.

NECESSARY AND SUFFICIENT CONDITION FOR CONVERGENCE

2. *Theorem*: *For the infinite product*

$$a_1 a_2 a_3 \ldots$$

to converge, it is necessary and sufficient that for every $\varepsilon > 0$ an $N > 0$ exist such that, for $n > N$ and for any positive integer p,

$$|a_{n+1} a_{n+2} \ldots a_{n+p} - 1| < \varepsilon.$$

Proof: (a) *Necessity* - Suppose that $a_n \neq 0$ for $n \geq m$ and that

$$a_m, \; a_m a_{m+1}, \; a_m a_{m+1} a_{m+2}, \; \ldots$$

converges to a limit $\underline{a} \neq 0$.

Let $\epsilon > 0$ be preassigned. Let $N > m$ be such that for $n > N$,
$$|a_m a_{m+1} \ldots a_n| > \frac{|a|}{2}$$
and
$$|a_m a_{m+1} \ldots a_{n+p} - a_m a_{m+1} \ldots a_n| < \frac{|a|}{2} \epsilon$$
for $p = 1, 2, 3, \ldots$.

Now
$$|a_m \ldots a_{n+p} - a_m \ldots a_n| = |a_m \ldots a_n| \, |a_{n+1} \ldots a_{n+p} - 1|.$$
Hence
$$|a_{n+1} \ldots a_{n+p} - 1| < \frac{\frac{|a|}{2}\epsilon}{\frac{|a|}{2}} = \epsilon.$$

Corollary: If $a_1 \ldots a_n \ldots$ converges then a_n approaches unity as n increases indefinitely.

(B) *Sufficiency:-* Let the condition be satisfied. Then, because a tends toward unity as n increases, at most a finite number of a's are zero. Suppose that $a_n \neq 0$ for $n \geq m$. Consider the sequence
$$a_m, \ a_m a_{m+1}, \ a_m a_{m+1} a_{m+2}, \ \ldots.$$
Let $M > m$ be such that
$$|a_{n+1} \ldots a_{n+p} - 1| < \frac{1}{2}$$
for $n \geq M$, $p = 1, 2, \ldots$.
Let
$$A = |a_m \ldots a_M|.$$
Then $A > 0$.

We know that for $n > M$,
$$|a_{M+1} a_{M+2} \ldots a_n - 1| < \frac{1}{2}.$$
Hence, multiplying by $|a_m \ldots a_M|$, we find
$$|a_m \ldots a_n - a_m \ldots a_M| < \frac{|a_m \ldots a_M|}{2}.$$
Then, for $n > M$,
$$\frac{A}{2} < |a_m \ldots a_n| < \frac{3A}{2}.$$

Let any $\epsilon > 0$ be assigned. Take $N > M$ in such a way that for $n > N$,
$$|a_{n+1} \ldots a_{n+p} - 1| < \epsilon$$
for $p = 1, 2, 3, \ldots$.

Then, for $n > N$,
$$|a_m \ldots a_{n+p} - a_m \ldots a_n| = |a_m \ldots a_n| \cdot |a_{n+1} \ldots a_{n+p} - 1| < \frac{3A\epsilon}{2}$$

Thus the sequence
$$a_m, a_m a_{m+1}, \ldots$$
converges. As $|a_m \ldots a_n| > \frac{A}{2}$ for $n > M$, the limit of this sequence is not zero.

Q.E.D.

ABSOLUTE CONVERGENCE

3. From now on, we write our infinite product in the form
$$(2) \quad (1 + u_1)(1 + u_2) \ldots (1 + u_n) \ldots$$
(The symbol $\prod_{n=1}^{\infty}(1 + u_n)$ is frequently used.)

For the product to converge, it is, of course, necessary that u_n tend toward zero as n increases indefinitely. It will be seen that this condition is not sufficient.

If the product
$$(3) \quad (1 + |u_1|)(1 + |u_2|) \ldots (1 + |u_n|) \ldots$$
converges, then (2) does also. For
$$|(1 + |u_n|) \ldots (1 + |u_{n+p}|) - 1| \geq |(1 + u_n) \ldots (1 + u_{n+p}) - 1|.$$

If (3) converges, we shall call (2) *absolutely convergent*.

Theorem: For
$$(4) \quad (1 + u_1) \ldots (1 + u_n) \ldots$$
to converge absolutely, it is necessary and sufficient that
$$(5) \quad u_1 + u_2 + \ldots + u_n + \ldots$$
converge absolutely.

Proof: (a) *Necessity* - Let $\varepsilon > 0$ be preassigned. Suppose that for $n > N$,
$$|(1 + |u_n|) \ldots (1 + |u_{n+p}|) - 1| < \varepsilon.$$
Now
$$(1 + |u_n|) \ldots (1 + |u_{n+p}|) = 1 + |u_n| + \ldots + |u_{n+p}| + |u_n||u_{n+1}| + \ldots + |u_n| \ldots |u_{n+p}|.$$
Hence
$$|(1 + |u_n|) \ldots (1 + |u_{n+p}|) - 1| \geq |u_n| + \ldots + |u_{n+p}|$$
and we have
$$|u_n| + \ldots + |u_{n+p}| < \varepsilon$$
for $n > N$. Thus (5) is absolutely convergent.

Q.E.D.

(b) *Sufficiency* - Suppose that
$$|u_1| + |u_2| + \ldots + |u_n| + \ldots$$
converges. Let an $\varepsilon > 0$ be preassigned and let N be such that
$$|u_{n+1}| + \ldots + |u_{n+p}| < \varepsilon$$
for $n > N$, $p = 1, 2, \ldots$.

Now, for any n,
$$1 + |u_n| \leq e^{|u_n|}$$

Thus
$$(1 + |u_{n+1}|) \cdots (1 + |u_{n+p}|) - 1 \leq e^{|u_{n+1}| + \cdots + |u_{n+p}|} - 1 < e^{\epsilon} - 1.$$

Now, if ϵ is very small, $e^{\epsilon} - 1$ is very small.

Q.E.D.

Note: It can be proved that the value of an absolutely convergent infinite product is not affected by a rearrangement of its terms.

INFINITE PRODUCTS OF ANALYTIC FUNCTIONS

4. Let
$$u_1(z), u_2(z), \ldots, u_n(z), \ldots$$
be a sequence of functions, all analytic in an open region A. Let

(6) $$[1 + u_1(z)] \cdots [1 + u_n(z)] \cdots$$

converge throughout A to a limit $V(z)$. We shall say that (6) converges *uniformly* to $V(z)$ in A if the convergence of the sequence
$$1 + u_1(z), [1 + u_1(z)][1 + u_2(z)], \ldots$$
to $V(z)$ is uniform in A.

Of course, if the convergence is uniform, $V(z)$ will be analytic in A.

Theorem: If there exists a convergent series of positive numbers
$$M_1 + M_2 + \cdots + M_n + \cdots$$
such that $|u_n(z)| < M_n$ throughout A, then
$$[1 + u_1(z)] \cdots [1 + u_n(z)] \cdots$$
converges uniformly in A, the convergence being absolute at every point of A.

Proof: That the product converges absolutely at every point of A is obvious. We need only attend to the question of uniformity.

The uniformity follows from the fact that for any z in A,
$$|[1 + u_1(z)] \cdots [1 + u_{n+p}(z)] - [1 + u_1(z)] \cdots [1 + u_n(z)]|$$
$$= |[1 + u_1(z)] \cdots [1 + u_n(z)]| \cdot |([1 + u_{n+1}(z)] \cdots [1 + u_{n+p}(z)] - 1)|$$
$$\leq (1 + M_1) \cdots (1 + M_n) \cdot [(1 + M_{n+1}) \cdots (1 + M_{n+p}) - 1]$$
$$= (1 + M_1) \cdots (1 + M_{n+p}) - (1 + M_1) \cdots (1 + M_n).$$

Now the product $(1 + M_1) \cdots (1 + M_n) \cdots$ is convergent, so that
$$(1 + M_1) \cdots (1 + M_{n+p}) - (1 + M_1) \cdots (1 + M_n)$$
is small when n is large.

XXXIX

The Weierstrass Factorization Theorem

INTEGRAL FUNCTIONS WITH A FINITE NUMBER OF ZEROS

1. Let it be required to determine all integral functions which have, at the points

$$a_1, a_2, \ldots, a_n,$$

distinct from one another, zeros of orders

$$p_1, p_2, \ldots, p_n,$$

respectively, and which have no other zeros.

One such function is

$$f(z) = (z - a_1)^{p_1} \ldots (z - a_n)^{p_n}$$

Consider any other such function, $F(z)$. We see that

$$\frac{F(z)}{f(z)}$$

is analytic all over the plane, even at the zeros of $f(z)$. Also, $F(z)/f(z)$ has no zeros. That is, $F(z)/f(z)$ is an integral function devoid of zeros. Now any such function is of the form $e^{g(z)}$, with $g(z)$ an integral function. Then

$$F(z) = e^{g(z)} (z - a_1)^{p_1} \ldots (z - a_n)^{p_n}.$$

Furthermore, any $F(z)$ as just written has the indicated zeros and no others.

ON INTEGRAL FUNCTIONS WITH AN INFINITE NUMBER OF ZEROS

2. Instead of saying, "$f(z)$ has p zeros at $\underline{a}$," we shall say, "$f(z)$ has zeros at $\underbrace{a, \ldots, a}_{p \text{ times}}$."

Let $f(z)$ be a non-constant integral function with an infinite number of zeros.

For $0 \leq |z| \leq 1$, there are at most a finite number of zeros. If there are any, we arrange them in a sequence, in the order of increasing moduli. If several zeros have equal moduli, we arrange them in any order.

We now adjoin to this sequence the zeros for which $1 < |z| \leq 2$, arranging them in the order of increasing moduli. Continuing in this fashion, we have all of the zeros of $f(z)$ written in an infinite sequence

$$a_1, a_2, \ldots, a_n, \ldots$$

with $|a_{n+1}| \geq |a_n|$ and with $\lim_{n \to \infty} |a_n| = \infty$.

CONSTRUCTION OF A FUNCTION WITH ASSIGNED ZEROS

3. Let

$$a_1, a_2, \ldots, a_n, \ldots$$

be any infinite sequence of complex numbers with
$$|a_{n+1}| \geq |a_n|$$
and with
$$\lim_{n \to \infty} |a_n| = \infty.$$

We propose to construct an integral function which vanishes for the points of this sequence and for no other points.

We assume, for the present, that no a_n is zero. This assumption will be removed later.

Suppose first that the series
$$\frac{1}{|a_1|} + \frac{1}{|a_2|} + \ldots + \frac{1}{|a_n|} + \ldots$$
converges. Then the function
$$f(z) = (1 - \frac{z}{a_1})(1 - \frac{z}{a_2}) \ldots (1 - \frac{z}{a_n}) \ldots$$
fulfills our requirements and every function of the type described is of the form
$$e^{g(z)}(1 - \frac{z}{a_1}) \ldots (1 - \frac{z}{a_n}) \ldots$$
with $g(z)$ an integral function.

But when $\Sigma\, 1/|a_n|$ diverges, the product $\pi(1 - z/a_n)$ will not as a rule converge.

We meet this situation as follows:

We choose positive integers
$$p_1, \ldots, p_n, \ldots$$
in such a way that for every $r > 0$,
$$|\frac{r}{a_1}|^{p_1} + \ldots + |\frac{r}{a_n}|^{p_n} + \ldots$$
converges.

For instance, we can take $p_n = n$. In that case, since for any r, $|a_n| > 2r$ for n sufficiently large, the nth term of the series has, for n large, a value less than $1/2^n$. Thus the series will surely converge.

We now consider the product

$$(1) \qquad \prod_{n=1}^{\infty} (1 - \frac{z}{a_n}) e^{\frac{z}{a_n} + \frac{1}{2}(\frac{z}{a_n})^2 + \ldots + \frac{1}{p_n - 1}(\frac{z}{a_n})^{p_n - 1}}$$

In this expression if some p_n is unity, the exponential is supposed to be absent from the nth factor, and the nth factor is simply $(1 - \frac{z}{a_n})$.

We shall prove that the product converges absolutely and uniformly in every bounded domain. By this we mean, firstly, that the product converges for every z; secondly, that the convergence is uniform in every bounded domain; finally, that if the nth factor in the product is written as
$$1 + u_n(z),$$
then $\pi[1 + |u_n(z)|]$ converges for every z.

When the uniform convergence is established, we shall know that the product converges to an integral function which vanishes only at the points at which its factors vanish. Now, since the exponentials vanish nowhere, the integral function will vanish at the points of our sequence and nowhere else.

CONVERGENCE PROOF

4. *Lemma: For $|u| < 1$, we have*

$$|e^u - 1| \leq 2|u|.$$

Proof:

$$|e^u - 1| = |u + \frac{u^2}{2!} + \frac{u^3}{3!} + \ldots| = |u| \cdot |1 + \frac{u}{2!} + \frac{u^2}{3!} + \ldots| \leq |u|(1 + \frac{1}{2!} + \frac{1}{3!} + \ldots) \leq 2|u|.$$

Q.E.D.

Consider now a circle with center at the origin and of radius r, where r is any positive number, arbitrarily large. Let m be such that $|a_n| > r$ for $n \geq m$.

Then, for $n \geq m$ and for $|z| < r$, we have

$$1 - \frac{z}{a_n} \neq 0.$$

Hence, $\log(1 - \frac{z}{a_n})$ is analytic for $|z| < r$. Taking that branch of the logarithm which is zero for $z = 0$, we have for $|z| < r$,

$$\text{Log}(1 - \frac{z}{a_n}) = -\frac{z}{a_n} - \frac{1}{2}(\frac{z}{a_n})^2 - \frac{1}{3}(\frac{z}{a_n})^3 - \ldots.$$

Then, for $|z| < r$,

$$1 - \frac{z}{a_n} = e^{-\frac{z}{a_n} - \frac{1}{2}(\frac{z}{a_n})^2 - \frac{1}{3}(\frac{z}{a_n})^3 - \ldots}$$

Thus, representing by v_n the nth factor of the product (1), we have for $|z| < r$,

$$v_n = e^{-[\frac{1}{p_n}(\frac{z}{a_n})^{p_n} + \frac{1}{p_n+1}(\frac{z}{a_n})^{p_n+1} + \ldots]}.$$

Now, for $|z| < \frac{r}{2}$, since $|a_n| > r$,

$$|\frac{1}{p_n}(\frac{z}{a_n})^{p_n} + \frac{1}{p_n+1}(\frac{z}{a_n})^{p_n+1} + \ldots| = |\frac{z}{a_n}|^{p_n} \cdot |\frac{1}{p_n} + \frac{1}{p_n+1}\frac{z}{a_n} + \ldots|$$

$$\leq |\frac{z}{a_n}|^{p_n} \cdot (1 + \frac{1}{2} + \frac{1}{4} + \ldots)$$

$$= 2|\frac{z}{a_n}|^{p_n}.$$

(*Note:* We have used the fact that $p_n \geq 1$ and that $|\frac{z}{a_n}| < \frac{1}{2}$.)

Hence, for $|z| < \frac{r}{2}$, by the lemma above,

$$|v_n - 1| \leq 4|\frac{z}{a_n}|^{p_n} < 4|\frac{r}{2a_n}|^{p_n}.$$

Thus, since $\Sigma|\frac{r}{a_n}|^{p_n}$ converges, we know from a previous theorem that

$$\prod_{n=m}^{\infty} v_n$$

converges absolutely and uniformly for $|z| < \frac{r}{2}$. The same is hence true of

$$\prod_{n=1}^{\infty} v_n.$$

As $\frac{r}{2}$ may be taken arbitrarily large, our result is established.

If it is desired to have a function which vanishes m times at the origin, as well as at $a_1, a_2, \ldots$, one may take

$$f(z) = z^m \prod_{n=1}^{\infty} (1 - \frac{z}{a_n}) e^{\frac{z}{a_n} + \ldots + \frac{1}{p_n - 1}(\frac{z}{a_n})^{p_n - 1}}$$

Any other integral function which has the indicated zeros, and no others, is given by

$$e^{g(z)} f(z)$$

where $g(z)$ is an integral function.

XL

Meromorphic Functions and Mittag-Leffler's Theorem

DEFINITION OF MEROMORPHIC FUNCTION

1. Let $f(z)$ be analytic in an open region A, except at a set of points of A at which it has poles. Then $f(z)$ is said to be *meromorphic* in A.

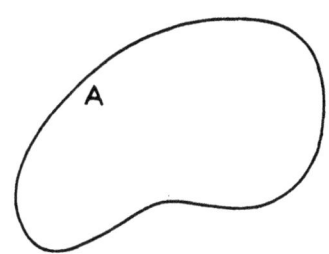

For instance, a rational function is meromorphic in the whole finite plane.

The poles need not actually exist. That is, if $f(z)$ is analytic in A, it is also meromorphic in A.

We note that, as poles are isolated singularities, $f(z)$ can have only a finite number of poles in any closed and bounded domain interior to A.

As a rule, when we speak of a "meromorphic function $f(z)$," without mentioning a domain, it is understood that $f(z)$ is meromorphic over the finite plane.

REPRESENTATION OF A MEROMORPHIC FUNCTION AS A QUOTIENT OF INTEGRAL FUNCTIONS

2. Let $f(z)$ be a function meromorphic all over the finite plane. Let $f(z)$ actually possess poles. The poles of $f(z)$ are isolated, so that, if they are infinite in number, they can be arranged in a sequence with ∞ as a limit.

Thus we can construct an integral function $g(z)$ which has, at every pole of $f(z)$, a zero whose order is that of the pole of $f(z)$.

Then $f(z) g(z)$ is an integral function. Let $f(z) g(z) = h(z)$. Then

$$f(z) = \frac{h(z)}{g(z)},$$

that is, every meromorphic function is a quotient of two integral functions.

MITTAG-LEFFLER'S THEOREM

3. Consider a meromorphic function $f(z)$ which has an infinite number of poles. Let the poles be arranged in a sequence

(1) $\qquad a_1, a_2, \ldots, a_n, \ldots$

with $|a_{n+1}| \geq |a_n|$ and $\lim_{n \to \infty} |a_n| = \infty$. In this sequence, multiple poles are not to be repeated. That is, even if there is a pole of higher order than the first at a_n, we write a_n only once.

At the pole a_n, $f(z)$ has a Laurent expansion

$$\frac{c_{-p_n, n}}{(z - a_n)^{p_n}} + \frac{c_{-p_n+1, n}}{(z - a_n)^{p_n-1}} + \ldots + \frac{c_{-1, n}}{z - a_n} + c_{0, n} + c_{1, n}(z - a_n) + \ldots,$$

which is valid inside of a circle about a_n on the circumference of which lies that pole of $f(z)$, distinct from a_n, which is closest to a_n.

Let G_n denote the principal part of the above expansion. That is,

$$(2) \qquad G_n = \frac{c_{-p_n, n}}{(z - a_n)^{p_n}} + \ldots + \frac{c_{-1, n}}{z - a_n}.$$

Suppose now that given a sequence (1) and any system of functions G_n (each G_n arbitrary), we desire to construct a meromorphic function $f(z)$ which is analytic except at the points a_n, and which has, at each a_n, a pole of order p_n with principal part G_n.

A first trial would perhaps be to form the series

$$(3) \qquad G_1 + G_2 + \ldots + G_n + \ldots$$

Suppose that, given a circle of any radius $r > 0$, where r is arbitrarily large, the series obtained from (3) by rejecting those terms for which $|a_n| < r$ converges uniformly for $|z| < r$. For instance, suppose that for $n > m$ we have $|a_n| \geq r$ and that the series

$$(4) \qquad G_{m+1} + G_{m+2} + \ldots$$

converges uniformly for $|z| < r$. Then the series (4) gives a function analytic for $|z| \leq r$, so that the series (3) is meromorphic for $|z| \leq r$, with a pole at each a_n for which $n \leq m$, the principal part at a_n being G_n.

As it was understood that r may be taken arbitrarily large, the series (3) gives a function such as we seek.

But, as a rule, (3) will not have the convergence properties exacted above.

We proceed to treat this difficulty.

We assume that $a_1 \neq 0$. It will be easy to see that this involves no loss of generality.

The function G_n is analytic for $|z| < |a_n|$. Hence, it has a Taylor expansion

$$(5) \qquad d_{0,n} + d_{1,n} z + \ldots + d_{p,n} z^p + \ldots$$

valid for $|z| < |a_n|$.

Let

$$\varepsilon_1, \varepsilon_2, \ldots, \varepsilon_n, \ldots$$

be any sequence of positive numbers such that

$$\varepsilon_1 + \varepsilon_2 + \ldots + \varepsilon_n + \ldots$$

converges.

Since (5) converges uniformly to $G_n(z)$ in any circle interior to the circle $|z| = |a_n|$, we can find a segment of the series (5), that is, a polynomial

$$P_n(z) = d_{0,n} + d_{1,n} z + \ldots + d_{q_n, n} z^{q_n}$$

such that

$$|G_n(z) - P_n(z)| < \varepsilon_n \text{ for } |z| < \frac{|a_n|}{2}.$$

Consider now the infinite series

$$(6) \qquad \sum_{n=1}^{\infty} [G_n(z) - P_n(z)].$$

Given any $r > 0$, let the positive integer m be such that $|a_n| > 2r$ for $n > m$. Then $G_n - P_n$, with $n > m$, is analytic for $|z| < 2r$. Furthermore, for $|z| < r$, we will have, since $r < \frac{|a_n|}{2}$,

$$|G_n - P_n| < \varepsilon_n.$$

Hence the series

$$\sum_{n=m+1}^{\infty} [G_n(z) - P_n(z)]$$

converges uniformly for $|z| < r$ and thus represents a function analytic for $|z| < r$. Since

$$\sum_{n=1}^{m} [G_n(z) - P_n(z)]$$

is a rational function which has at a_n with $n \leq m$ a pole of principal part G_n, the series

$$\sum_{n=1}^{\infty} [G_n(z) - P_n(z)]$$

is meromorphic for $|z| < r$, with a pole at each a_n with $|a_n| < r$ at which the principal part is G_n.

As r may be taken arbitrarily large, (6) gives a function meromorphic all over the finite plane, with poles of the desired type.

If $a_1 = 0$, we take $P_1(z) = 0$ and the series (6) evidently gives the solution of our problem.

Let $f(z)$ be the function given by (6). Consider any other function $F(z)$ which is analytic except at the points a_n and which has at each a_n a pole of principal part G_n.

Then

$$F(z) - f(z)$$

is analytic even at the points a_n, for the principal part of $F(z)$ at a_n is removed by that of $f(z)$ when we subtract.

Then

$$F(z) = f(z) + g(z)$$

where $g(z)$ is an integral function.

XLI

Theory of Residues

RESIDUE AT A FINITE POINT

1. Let $f(z)$ have an isolated singularity at a finite point $\underline{a}$. Let the Laurent development be

$$c_0 + c_1(z - a) + c_2(z - a)^2 + \ldots$$
$$+ c_{-1}(z - a)^{-1} + c_{-2}(z - a)^{-2} + \ldots$$

The quantity c_{-1} is called the *residue* of $f(z)$ at $\underline{a}$. We see immediately that the residue is given by

$$\frac{1}{2\pi i} \int_C f(z)\, dz$$

where C is a sufficiently small positively sensed circle with center at $\underline{a}$.

RESIDUE AT INFINITY

2. Let $f(z)$ be analytic in a neighborhood of ∞, with a Laurent expansion at ∞

$$c_0 + c_1 z + c_2 z^2 + \ldots$$
$$+ c_{-1} z^{-1} + c_{-2} z^{-2} + \ldots$$

The quantity $-c_{-1}$ is called the residue of $f(z)$ at ∞. Thus, the residue is

$$\frac{1}{2\pi i} \int_C f(z)\, dz$$

where C is a large, *negatively* sensed circle.

FORMULA FOR RESIDUES WITHIN A CURVE

3. *Theorem: Let $f(z)$ be analytic, except for isolated singularities, in a simply connected open region A. Let C be a rectifiable Jordan curve in A, at every point of which $f(z)$ is analytic. Then the sum of the residues of $f(z)$ for the singularities within C is given by*

$$\frac{1}{2\pi i} \int_C f(z)\, dz,$$

where the integration is performed in the positive sense.

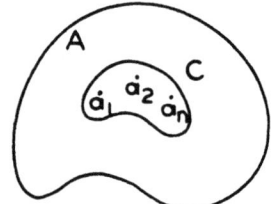

Proof: There can evidently be only a finite number of singularities within C. Let them be $a_1, \ldots, a_n$.

Place a small circle Γ_j about each point a_j. Then $f(z)$ is analytic in an open region containing $C, \Gamma_1, \ldots \Gamma_n$ and the open region bounded by them. Hence,

$$\int_C f(z)\, dz - \int_{\Gamma_1} f(z)\, dz - \ldots - \int_{\Gamma_n} f(z)\, dz = 0,$$

or

$$\frac{1}{2\pi i} \int_C f(z)\, dz = \sum_{j=1}^{n} \frac{1}{2\pi i} \int_{\Gamma_j} f(z)\, dz,$$

and the jth term in the second member is the residue at a_j.

Q.E.D.

THE RESIDUES OF A RATIONAL FUNCTION

4. Theorem: *The sum of the residues of a rational function is zero.*

Proof: Let C be a circle which contains in its interior all of the finite poles of a rational function $f(z)$. Then the sum of the residues at the finite poles is

$$\frac{1}{2\pi i} \int_C f(z)\, dz,$$

the integration being performed in the positive sense. The residue at ∞ is

$$\frac{1}{2\pi i} \int_C f(z)\, dz$$

the integration being performed in the negative sense. Thus the sum of the residues is zero.

Example: Let

$$f(z) = \frac{z}{z^2 - 1}.$$

The poles are $+1$ and -1. As

$$f(z) = \frac{z}{z+1} \cdot \frac{1}{z-1}$$

and as the Taylor expansion of $\frac{z}{z+1}$ at 1 is of the form

$$\tfrac{1}{2} + c_1(z-1) + \ldots,$$

the Laurent expansion of $f(z)$ at 1 is

$$\frac{1}{2(z-1)} + c_1 + c_2(z-1) + \ldots,$$

so that the residue at 1 is $\tfrac{1}{2}$.

Similarly, the residue at -1 is $\tfrac{1}{2}$.

Now for $|z| > 1$, we have

$$\frac{z}{z^2-1} = \frac{1}{z}\left(\frac{1}{1-\frac{1}{z^2}}\right) = \frac{1}{z}\left(1 + \frac{1}{z^2} + \frac{1}{z^4} + \ldots\right) = \frac{1}{z} + \frac{1}{z^3} + \ldots$$

so that the residue at ∞ is -1.

Thus, the sum of the residues is 0.

ON THE LOGARITHMIC DERIVATIVE OF AN ANALYTIC FUNCTION

5. By the *logarithmic derivative* of $f(z)$, we mean

$$\frac{f'(z)}{f(z)}.$$

Let $f(z)$, not identically zero, be analytic for a vicinity of a finite point $\underline{a}$, with a Laurent development which contains at most a finite number of negative powers. Let

$$f(z) = c_p(z-a)^p + c_{p+1}(z-a)^{p+1} +$$

with $c_p \neq 0$. The integer p may be positive, negative or zero. Then

$$f'(z) = pc_p(z-a)^{p-1} + (p+1)c_{p+1}(z-a)^p + \ldots.$$

165

By what we know of the quotient of two analytic functions which are analytic in a vicinity of a point, $f'(z)/f(z)$ is analytic for a vicinity of $\underline{a}$, with a Laurent expansion

$$\frac{f'(z)}{f(z)} = \frac{p}{z-a} + \text{higher powers.}$$

Thus, if p is zero, $\frac{f'(z)}{f(z)}$ is analytic at $\underline{a}$, while if p is positive or negative, $\frac{f'(z)}{f(z)}$ has a simple pole (pole of the first order) at $\underline{a}$, with residue p.

Thus:

1. *If $f(z)$ is analytic at $\underline{a}$ and distinct from zero at $\underline{a}$, $f'(z)/f(z)$ is analytic at $\underline{a}$.*

2. *If $f(z)$ has a zero of order m at a, $f'(z)/f(z)$ has a simple pole at a, with residue m.*

3. *If $f(z)$ has a pole of order m at a, $f'(z)/f(z)$ has a simple pole at a, with residue -m.*

The Fundamental Theorem of the Calculus of Residues

6. *Theorem: Let $f(z)$ be meromorphic in a simply connected open region A. Let $\Phi(z)$ be analytic in A. Let C be a rectifiable Jordan curve, lying in A, on which $f(z)$ is analytic, but nowhere zero. Let $f(z)$ have within C the zeros*

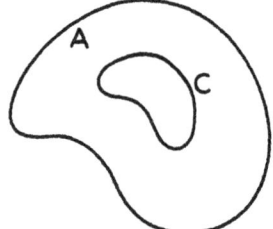

$$a_1, \ldots, a_r$$

of multiplicities

$$p_1, \ldots, p_r$$

respectively and the poles

$$b_1, \ldots, b_s$$

of multiplicities

$$q_1, \ldots, q_s$$

respectively. Then

$$\frac{1}{2\pi i} \int_C \Phi(z) \frac{f'(z)}{f(z)} dz = \sum_{j=1}^{r} p_j \Phi(a_j) - \sum_{j=1}^{s} q_j \Phi(b_j).$$

Proof: The value of

$$\frac{1}{2\pi i} \int_C \Phi(z) \frac{f'(z)}{f(z)} dz$$

is the sum of the residues of $[\Phi(z) f'(z)]/f(z)$ for the singularities of that function within C. (Note that the integrand is analytic along C.)

Let

$$g(z) = \Phi(z) \frac{f'(z)}{f(z)}$$

Then $g(z)$ can have a singularity only where $f(z)$ has a pole or a zero.

Consider a point a_j. The expansion of $f'(z)/f(z)$ at a_j is

$$\frac{p_j}{z - a_j} + \text{higher powers.}$$

Let the expansion of $\Phi(z)$ at a_j be

$$\Phi(a_j) + \Phi'(a_j)(z - a_j) + \ldots.$$

Then, if $\Phi(a_j) \neq 0$, $g(z)$ will have a simple pole at a_j, with residue $p\Phi(a_j)$. If $\Phi(a_j) = 0$, $g(z)$ is analytic at a_j.

Similarly, at a point b_j, $g(z)$ has a pole of the first order with residue $-q_j \Phi(b_j)$, or is analytic, according as $\Phi(b_j) \neq 0$ or $\Phi(b_j) = 0$.

From these considerations, it is plain that the sum of the residues of $g(z)$ within C is

$$\sum_{j=1}^{r} p_j \Phi(a_j) - \sum_{j=1}^{s} q_j \Phi(b_j).$$

This proves the theorem.

SPECIAL CASES

7. If we take $\Phi(z) = 1$, we secure the following theorems.

Theorem: Let $f(z)$ be meromorphic in a simply connected open region A. Let C be a rectifiable Jordan curve in A, along which $f(z)$ is analytic and nowhere zero. Then

$$\frac{1}{2\pi i} \int_C \frac{f'(z)}{f(z)} dz$$

is the difference between the number of zeros of $f(z)$ within C and the number of poles of $f(z)$ within C.

Theorem: Let $f(z)$ be analytic in a simply connected open region A. Let C be a rectifiable Jordan curve, lying in A, on which $f(z)$ is nowhere zero. Then the number of zeros of $f(z)$ within C is given by

$$\frac{1}{2\pi i} \int_C \frac{f'(z)}{f(z)} dz.$$

APPLICATION TO THE FUNDAMENTAL THEOREM OF ALGEBRA

8. Let $f(z)$ be a non-constant polynomial

$$a_0 z^n + \ldots + a_n$$

with $a_0 \neq 0$. We know that $|f(z)|$ is large when $|z|$ is large. Let C be a circle with center at the origin, outside of which $f(z)$ is not zero. We shall examine the Laurent expansion of $f'(z)/f(z)$ for the exterior of C.

As

$$f'(z) = na_0 z^{n-1} + \ldots + a_{n-1},$$

we see from the theorems on the quotient of two functions that the Laurent development of $f'(z)/f(z)$ for ∞ is

$$\frac{n}{z} + \text{lower powers}.$$

Thus, the value of

$$\frac{1}{2\pi i} \int_\Gamma \frac{f'(z)}{f(z)} dz,$$

where Γ is a circle containing C in its interior, is n.

Hence, *a polynomial of degree $n > 0$ has precisely n zeros.*

XLII

Certain Important Theorems

ROUCHÉ'S THEOREM

1. *Let $f(z)$ and $\varphi(z)$ be two functions, analytic in a simply connected open region A. Let C be a rectifiable Jordan curve lying in A. Suppose that along C, $f(z)$ is nowhere zero, and*

$$|\varphi(z)| < |f(z)|.$$

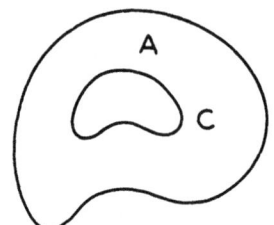

Then the function $f(z) + \varphi(z)$ has the same number of zeros within C as $f(z)$ has.

Proof: For any real value of t on the closed interval $[0, 1]$, the function

$$g_t(z) = f(z) + t\,\varphi(z)$$

is analytic in A and vanishes nowhere on C. Let N_t represent the number of zeros of $g_t(z)$. Then

$$(1) \qquad N_t = \frac{1}{2\pi i}\int_C \frac{g'_t(z)}{g_t(z)}\,dz = \frac{1}{2\pi i}\int_C \frac{f'(z) + t\,\varphi'(z)}{f(z) + t\,\varphi(z)}\,dz.$$

It is easy to prove that a small change in t produces a small change in the second integral written in (1), so that N_t is a continuous function of t. As N_t is an integer for every value of t, N_t is a constant. Thus, $N_1 = N_0$ and this proves the theorem.

APPLICATION TO THE FUNDAMENTAL THEOREM OF ALGEBRA

2. Let

$$f(z) = a_0 z^n + \ldots + a_n$$

with $n \geq 1$ and $a_0 \neq 0$. If C is a circle of large radius, with center at the origin, we have all along C,

$$|a_1 z^{n-1} + \ldots + a_n| < |a_0 z^n|.$$

Thus, $f(z)$ has the same number of zeros within C as $a_0 z^n$ does. That number is n.

Q.E.D.

PRESERVATION OF NEIGHBORHOODS

3. *Theorem: Let $f(z)$ be a non-constant function, analytic at a point a. Then, in any neighborhood of a, $f(z)$ assumes all values in some neighborhood of $f(a)$.*

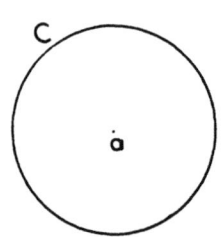

That is, if δ is any positive number such that $f(z)$ is analytic for $|z - a| < \delta$, there exists a positive η such that, given any number c with $|c - f(a)| < \eta$, there is a z with $|z - a| < \delta$ for which $f(z) = c$.

Proof: The places at which $f(z)$ assumes the value $f(a)$ cannot have a as a limit point, else $f(z)$ would be equal to the constant $f(a)$ everywhere.

168

Given an open region containing $\underline{a}$ in which $f(z)$ is analytic, let C be a circle with center at $\underline{a}$, lying with its interior in the open region, such that $\underline{a}$ is the only value of z within or on C for which $f(z) = f(a)$.

Then the function $f(z) - f(a)$ is nowhere zero on C. Let the lower bound of $|f(z) - f(a)|$ on C be η. Then $\eta > 0$.

Let $f(z)$ assume the value $f(a)$ precisely n times at $\underline{a}$. Then $f(z) - f(a)$ has n zeros within C. Let c be any number such that $|c - f(a)| < \eta$. By Rouché's theorem,

$$f(z) - c = [f(z) - f(a)] + [f(a) - c]$$

has precisely n zeros within C.

Q.E.D.

ON THE MAXIMUM OF THE MODULUS OF AN ANALYTIC FUNCTION

4. Let E be a bounded perfect set of points of the complex plane. Let $f(z)$ be defined on E and continuous at every point of E. It is easy to see that there is a point of E at which $|f(z)|$ attains a maximum value.

Theorem: Let A be a bounded open region. Let $f(z)$ be defined in A and upon the boundary of A, $f(z)$ being analytic in A and continuous upon the boundary. Then the maximum value of $|f(z)|$ is attained at a point on the boundary of A.

Remark: If $f(z)$ is not a constant, $|f(z)|$ attains its maximum value only upon the boundary.

Proof: The theorem is trivial for a constant $f(z)$. Suppose that $f(z)$ is not a constant. Suppose that there is a point $\underline{a}$ in A at which the modulus is a maximum. Since $f(z)$ is analytic at $\underline{a}$, the values of $f(z)$ for z close to $\underline{a}$ cover a neighborhood of $f(a)$. Thus, $|f(a)|$ cannot be a maximum.

ON THE MINIMUM OF THE MODULUS OF AN ANALYTIC FUNCTION

5. *Theorem: Let $f(z)$ be as in §4, and furthermore, let $f(z)$ be nowhere zero in A. Then $|f(z)|$ assumes its minimum value on the boundary of A.*

Proof: If $f(z)$ is zero somewhere on the boundary, we have our result. Suppose that $f(z)$ is nowhere zero on the boundary. Then $g(z) = \dfrac{1}{f(z)}$ satisfies the hypothesis of §4. Thus $|g(z)|$ is a maximum at a point on the boundary. Hence $|f(z)|$ attains its minimum on the boundary.

FUNDAMENTAL THEOREM OF ALGEBRA

6. Let

$$f(z) = a_0 z^n + \ldots + a_n$$

with $n \geq 1$, $a_0 \neq 0$. Suppose that $f(z)$ is nowhere zero.

Consider a large circle C with center at the origin. As $|f(z)|$ is very large on C, $|f(z)|$ cannot attain its minimum on C. This contradiction proves that $f(z)$ has zeros.

SEQUENCE OF ANALYTIC FUNCTIONS

7. Let

$$u_1(z), \ldots, u_n(z), \ldots$$

be a sequence of functions, each analytic in the interior of a rectifiable Jordan curve C and continuous on C. Let the sequence converge uniformly on C. Then the sequence converges uniformly in the domain composed of C and its interior. (Diagram, top of page 170.)

Proof: Let $\varepsilon > 0$ be assigned. Let N be such that for $n > N$, $p = 1, 2, \ldots$,

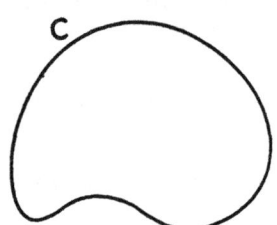

(2) $$|u_{n+p}(z) - u_n(z)| < \varepsilon$$

for every point of C.

Consider any particular $n > N$ and any particular p. The modulus of the function $u_{n+p}(z) - u_n(z)$ attains its maximum on C. Hence, (2) holds in the interior of C as well as on C.

Q.E.D.

XLIII

Variation of the Amplitude of a Continuous Function along a Continuous Curve

VARIATION OF AMPLITUDE

1. We consider a continuous curve C, defined by

$$z = \varphi(t) + i\psi(t) \qquad a \leq t \leq b.$$

Let $f(z)$ be a function of z, defined, and continuous, along C. We assume that $f(z)$ is distinct from zero at every point of C.

We recall that $f(z)$ is a continuous function $F(t)$ of t.

At any point z_0 of C, the complex number $f(z_0)$ has an infinite number of amplitudes, which differ from each other by multiples of 2π. We recall that if $a_0 + ib_0$ is any one of the logarithms of $f(z_0)$, then b_0 is an amplitude of $f(z_0)$. This is because

$$f(z_0) = e^{a_0 + ib_0} = e^{a_0} e^{ib_0}$$
$$= e^{a_0}(\cos b_0 + i \sin b_0).$$

We are going to prove the existence of a function <u>amp $f(z)$</u>, continuous along C, whose value for each z is an amplitude of $f(z)$. There will prove to be an infinite number of continuous functions of this type, any one of which differs from any other all along C by a constant integral multiple of 2π.

Proof: Since $f(z)$ is continuous and distinct from zero along C, $|f(z)|$ is a real positive function, continuous along C. Since C is a closed and bounded set, $|f(z)|$ assumes a minimum somewhere on C. This minimum, call it m, is positive.

Let $\delta > 0$ be such that for $|t_2 - t_1| < \delta$,

$$|f(z_2) - f(z_1)| < \frac{m}{2},$$

where $z_1 = \varphi(t_1) + i\psi(t_1)$, $z_2 = \varphi(t_2) + i\psi(t_2)$. Let (a, b) be divided by points t_i such that

$$t_0 = a,\ t_1 > t_0,\ t_2 > t_1,\ \ldots,\ t_n = b,$$

with $t_{i+1} - t_i < \delta$, $i = 0, \ldots, n - 1$.

Let the corresponding points on the curve be

$$z_0,\ z_1,\ \ldots,\ z_n.$$

We shall now use a new complex variable w. Consider a circle in the w-plane, with center at $f(z_0)$ and radius $\frac{m}{2}$. Since $|f(z_0)| \geq m$, the origin is outside the circle. Hence, $\log w$ is analytic within the circle. Of course, there are an infinite number of such analytic functions $\log w$, differing from one another by multiples of $2\pi i$. The coefficients of i in their values at z_0 are the various amplitudes of $f(z_0)$.

Let us select a definite amplitude of $f(z_0)$, say θ, and then choose the function log w so that its coefficient of i at $f(z_0)$ is θ.

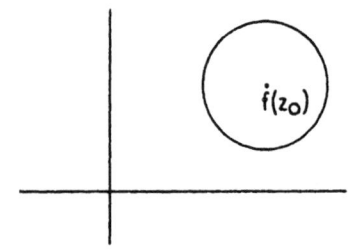

The coefficient of i in log w is certainly a continuous function of w within the circle.

For each t in (t_0, t_1), we use for amp $f(z)$ the coefficient of i in log w, where $w = f(z)$. That is, for every t, we take the corresponding z on the curve and use the value $f(z)$ for w. For t on (t_0, t_1), $f(z)$ will be within the circle used here.

Then amp $f(z)$ is a continuous function of t in (t_0, t_1).

Let θ_1 be the value of amp $f(z)$, as thus defined, for $t = t_1$.

We take a circle with $f(z_1)$ as center and with $\frac{m}{2}$ for radius. In this circle, we take an analytic log w whose coefficient of i at $f(z_1)$ is θ_1. Then, if we define amp $f(z)$ in (t_1, t_2) as equal to the coefficient of i in log w, where $w = f(z)$, amp $f(z)$ will be continuous in (t_1, t_2).

In all, we have a function amp $f(z)$, defined and continuous throughout (t_0, t_2).

Proceeding step by step, we secure an amp $f(z)$, defined and continuous throughout (a, b).

Any other continuous function which, at each point of C has a value which is an amplitude of $f(z)$ must differ from amp $f(z)$ as above defined, at each point, by an integral multiple of 2π. As the difference is continuous, it must be constant, and a multiple of 2π.

2. Take any one of the infinite set of continuous functions amp $f(z)$. Let us denote it, simply, by amp $f(z)$.

By the variation of the amplitude of $f(z)$ along C, we shall mean

$$\text{amp } f(Q) - \text{amp } f(P)$$

where P and Q are, respectively, the first and last points of the curve. This difference is the same for all possible functions amp $f(z)$ obtained as above from $f(z)$. (Note that two such functions have a constant difference.) If the curve is closed, the variation of the amplitude of $f(z)$ along the curve does not depend on the choice of the first point of the curve.

3. Consider a Jordan curve

$$z = \varphi(t) + i\psi(t) \qquad \alpha \leq t \leq \beta.$$

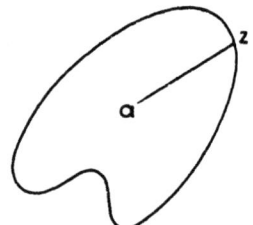

Let $\underline{a}$ be any point within the curve. It is geometrically plausible that if the curve is positively sensed, the variation of the amplitude of $z - a$ long the curve is 2π and that if the curve is negatively sensed, the variation of the amplitude is -2π. We shall assume these results, which are rigorously proved in the theory of analysis situs.

ON THE ZEROS OF AN ANALYTIC FUNCTION

4. Let A be a simply connected open region. Let C be a positively sensed rectifiable Jordan curve in A. Let $f(z)$ be a function analytic in A, whose only zeros are within C.

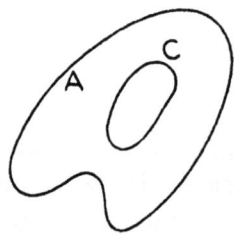

We are going to prove that the *number of zeros of* $f(z)$ *within* C *is equal to the variation of the amplitude of* $f(z)$ *along* C *divided by* 2π.

Proof: We consider first the case in which $f(z)$ has no zeros within C.

In that case, $\int_C \frac{f'(z)}{f(z)} dz = 0$. Also, we know that log $f(z)$ is

172

analytic in A. Now the integral of $\frac{f'(z)}{f(z)}$ between any two points P and Q of A is

$$\log f(Q) - \log f(P),$$

which equals

$$[\log |f(Q)| + i \text{ amp } f(Q)] - [\log |f(P)| + i \text{ amp } f(P)]$$

or

$$\log |f(Q)| - \log |f(P)| + i [\text{amp } f(Q) - \text{amp } f(P)].$$

Now, if P and Q coincide, this result is

$$i [\text{amp } f(Q) - \text{amp } f(P)],$$

or i times the variation of the amplitude of $f(z)$ along C.

Hence, since $\int_C \frac{f'(z)}{f(z)} dz = 0$, the variation of amp $f(z)$ along C is zero.

Suppose now that $f(z)$ has zeros at $a_1, \ldots, a_n$ within C.

Then

$$\frac{f(z)}{(z - a_1) \ldots (z - a_n)}$$

has no zeros within C, so that the variation of its amplitude along C is zero.

Now

$$\text{amp } \frac{f(z)}{(z - a_1) \ldots (z - a_n)} = \text{amp } f(z) - [\text{amp } (z - a_1) + \ldots + \text{amp } (z - a_n)]$$

Hence, the variation of amp $f(z)$ along C is the sum of the variations of amp $(z - a_1), \ldots,$ amp $(z - a_n)$. But as each a_j is within C, each amp $(z - a_j)$ increases by 2π as z moves around C.

Thus the variation of amp $f(z)$ along C is $2\pi n$.

Q.E.D.

XLIV

The Functions $\sqrt[n]{z}$, log z

THE FUNCTION $\sqrt{z}$

1. The equation

$$(1) \qquad w^2 = z$$

has two distinct roots when $z \neq 0$ and a single root when $z = 0$.

We shall call the symbol $\sqrt{z}$ (or $z^{\frac{1}{2}}$), a *two-valued function of* z and we shall call the solutions of (1) for any value of z, the *values* of $\sqrt{z}$ for that value of z.

The two values of $\sqrt{z}$ are negatives of each other.

The modulus of either value is $\sqrt{|z|}$.

The double of any amplitude of a value of $\sqrt{z}$ must be an amplitude of z.

To find amplitudes of the two values of $\sqrt{z}$, we can take any amplitude θ of z, and then

$$\frac{\theta}{2}, \frac{\theta}{2} + \pi$$

will be amplitudes for the two mentioned values. (Note that $2(\frac{\theta}{2} + \pi) = \theta + 2\pi$ and that $\theta + 2\pi$ is an amplitude of z.)

2. Consider any simply connected open region A which does not contain the origin.

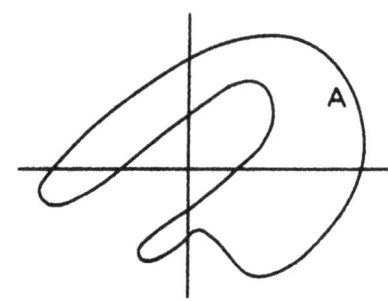

We shall show that *there exist in* A, *two analytic functions, the square of either of which is* z. *The two analytic functions are negatives of each other.* Any function, analytic in A, whose square is z, coincides with one or the other of the above two functions.

Proof: Since A is simply connected and z is not zero at any point of A, log z is analytic in A. Let any branch of log z be selected, that is, any function analytic in A whose exponential is z.

Then

$$e^{\frac{1}{2} \log z}$$

is analytic in A. Now

$$(e^{\frac{1}{2} \log z})^2 = e^{\log z} = z.$$

Hence, the two functions

$$e^{\frac{1}{2} \log z}, -e^{\frac{1}{2} \log z}$$

answer our requirements. We call these functions the *branches of* $\sqrt{z}$ *in* A.

If a third function, analytic in A, has z for its square, it certainly coincides with one or the other of the above two functions on some infinite set of points with a limit point in A. Hence it coincides throughout A with one of the two branches of $\sqrt{z}$.

3. We shall prove that there is no function analytic in a neighborhood of the origin whose square is z. Suppose that such a function, $g(z)$, exists. Then $g(z)$ must have a zero at the origin. Then $z = [g(z)]^2$ has a zero of order at least 2 at the origin. This proves our statement.

4. Consider any continuous curve C, not passing through the origin, which joins two points a and b, which may or may not coincide.

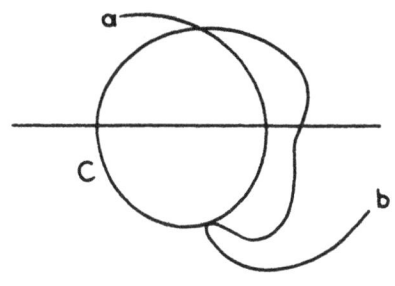

We shall prove the *existence of two functions, negatives of each other, continuous along* C, *the square of either of which is* z.

The function $|z|$ is continuous on C.

Hence, $\sqrt{|z|}$ is continuous on C. (We understand that $\sqrt{|z|} > 0$.)

Let any value of amp z be selected at a.

This determines uniquely a function amp z continuous on C.

Then the function g(z) defined by

$$g(z) = \sqrt{|z|}\, e^{\frac{1}{2} i\, \text{amp}\, z},$$

that is, by

$$g(z) = \sqrt{|z|}\, (\cos \tfrac{1}{2} \text{amp}\, z + i \sin \tfrac{1}{2} \text{amp}\, z)$$

is continuous on C and its square is z. The function $-g(z)$ also answers our requirements. It is easy to prove that $g(z)$ and $-g(z)$ are the only two functions continuous on C, whose squares are z.

Suppose that C is closed, that is, that b coincides with a. If the variation of amp z along C is an even multiple of 2π, that is, speaking geometrically, if C makes an even number of turns about the origin, the value of $g(z)$ at b (the terminal point of C) is the same as the value at a (initial point of C). For $\frac{1}{2}$ amp z will increase by a multiple of 2π, so that $e^{\frac{1}{2} i\, \text{amp}\, z}$ will not change.

On the other hand, if the variation of amp z is an odd multiple of 2π, then $\frac{1}{2}$ amp z will increase by an odd multiple of π, so that the value of $g(z)$ at b is the negative of the value of a.

Speaking intuitively, if $\sqrt{z}$ is followed along a closed curve, starting from some point on the curve, one returns to the starting point with the same value of $\sqrt{z}$ if the curve winds about the origin an even number of times, and one returns with the negative of the original value if the curve makes an odd number of turns about the origin.

5. The foregoing results tell, in an arithmetic way, everything which is essential about the behavior of $\sqrt{z}$. We shall now introduce, in an intuitive way, the idea of *Riemann Surface*, which will furnish us with an excellent physical picture of the behavior of $\sqrt{z}$.

Let a plane, representing the complex plane, be cut along the positive real axis. That is, let the positive real axis be deleted from the plane. There is left an open region in which both branches of $\sqrt{z}$ are analytic.

Now suppose that we have two planes like the above cut along the positive real axis.

Suppose that we label the points of one of the dissected planes with the values of one of the branches of $\sqrt{z}$ and label the points of the second dissected plane with the values of the other branch.

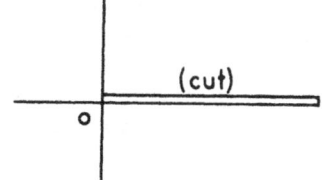

As to the cut itself, we can imagine any point of it as belonging either to the upper edge of the cut or to the lower edge. (This is intuitive. The arithmetic has been disposed of in §§1-4.) In each plane, we attribute to $\sqrt{z}$, at a point on the upper edge of the cut, that value which is the limit approached by $\sqrt{z}$ when one approaches the point on the cut from the upper half of the plane. For the lower edge of a cut, one proceeds similarly. The two values of a branch of $\sqrt{z}$ at coincident points on opposite sides of a cut are negatives of each other.

Now we take the two dissected planes and fuse the upper edge of the cut for one plane to the lower edge of the cut for the second plane. We fuse the lower edge of the first plane with the upper edge of the second. We form in this way a two-sheeted surface to each point of which one value of $\sqrt{z}$ is attached.

It is clear that if we start from any point of the surface, and follow a curve which is closed but winds about the origin, we will return to the same point on the surface if the curve winds about the origin an even number of times, but that we will reach the point on the other sheet of the surface which has the same z as the starting point if one winds about the origin an odd number of times.

Thus, the two-sheeted surface, labeled with the values of $\sqrt{z}$, gives a clear picture of the behavior of the function $\sqrt{z}$. We call the surface the *Riemann surface* of $\sqrt{z}$.

6. For $\sqrt[n]{z}$, where n is any positive integer greater than 1, we obtain similarly an n-sheeted surface.

THE FUNCTION LOG z

7. The function log z can be treated in the same way. If we follow log z along a curve which winds n times about the origin, in the positive sense, log z increases by $2n\pi i$.

We therefore take the complex plane and cut it along the positive real axis. In the simply connected open region A which results, log z is analytic.

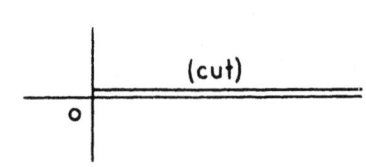

Consider any branch of log z in A. Its two values at points on the cut on opposite edges of the cut differ by $2\pi i$.

Consequently, if we take a second such plane and label its points with values greater by $2\pi i$ than the values at corresponding points in the first such plane, and if we fuse the upper edge of the first plane to the lower edge of the second plane, we will have a two-sheeted surface which will give a clear picture of two branches of log z. To visualize all of the branches of log z, we can use an infinite number of planes, each of which we associate with one of the integers

$$\ldots, -4, -3, -2, -1, 0, 1, 2, 3, 4, \ldots$$

Consider the sheet associated with any integer n. We attribute to each point of it, a value of log z which exceeds by $2n\pi i$ the value at the corresponding point of the first sheet.

If we fuse the lower edge of the nth sheet to the upper edge of the (n + 1)st sheet, we will have a Riemann surface of an infinite number of sheets, to each point of which is attached one value of log z.

8. In each of the Riemann surfaces considered above, the point on the surface for which z = 0 is called a *branch point* of the surface.

Also, we adjoin to each of the surfaces an ideal point ∞, which is also called a branch point of the surface. The reason for this is that if we move around a large circle with center at the origin, the values of the functions change. (Note that a change results in the case of *every* circle. The reason we refer to a large circle is that it is the large circles, in other cases, which decide whether or not ∞ is a branch point.)

XLV

Analytic Continuation

ELEMENTS AND THEIR CONTINUATIONS

1. By an *analytic element*, we shall mean a power series

$$c_0 + c_1(z - a) + \ldots + c_n(z - a)^n + \ldots,$$

with a radius of convergence greater than 0. The point a will be called the *center* of the element.

We shall represent the above element by the symbol $P(z; a)$.

2. Consider the analytic function $f(z)$ which the above element defines within the circle of convergence of the element. Let a' be any point within the circle. We can develop $f(z)$ in powers of $z - a'$, obtaining a second analytic element

$$c_0' + c_1'(z - a') + \ldots + c_n'(z - a')^n + \ldots$$

whose radius of convergence is at least as great as the distance from a' to the first circle. On the other hand, the second circle may extend outside the first one.

But certainly, in the open region which is common to the interiors of the two circles, the power series converge to the same values.

We shall call the second analytic element an *immediate continuation* of the first one. Every analytic element has an infinite number of immediate continuations, one for every point within its circle of convergence.

Let us give an example to show that the second power series may converge for points outside the first circle.

Consider the function $\frac{1}{z}$, which has a simple pole for $z = 0$. The expansion in powers of $z - a$, for any $a \neq 0$, has a radius of convergence $|a|$. For $\frac{1}{z}$ is analytic inside the circle $|z - a| = |a|$, on which the origin lies, but is not analytic in any larger circle with a as center. Evidently, if a' is any point within the above circle, not on the line joining a to 0, the expansion about a' will converge for certain points outside that circle.

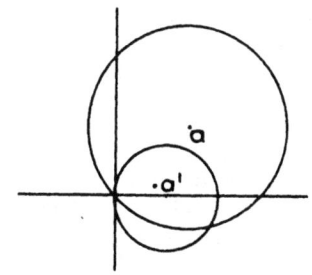

3. Each immediate continuation of an analytic element has immediate continuations, which in turn have their immediate continuations. etc.

By a *chain* of analytic elements, we shall mean a finite sequence of elements

$$P(z; a), P(z; a'), P(z; a''), \ldots, P(z; a^{(n)}),$$

each element after the first being an immediate continuation of its predecessor.

We shall call an analytic element $P(z; b)$ a *continuation* (not necessarily immediate) of an element $P(z; a)$ if there exists a chain whose first element is $P(z; a)$ and whose last element is $P(z; b)$.

It is not hard to prove that if $P(z; b)$ is a continuation of $P(z; a)$, then $P(z; a)$ is a continuation of $P(z; b)$.

177

It is easy to see that if $P(z; b)$ and $P(z; c)$ are continuations of $P(z; a)$, they are continuations of each other.

THE MONOGENIC ANALYTIC FUNCTION

4. Consider any analytic element $P(z; a)$ and the totality of analytic elements which are continuations of it. Weierstrass called the totality of those elements a *monogenic analytic function*. Thus an analytic function for Weierstrass was a collection of power series.

We observe that any element which is a continuation of $P(z; a)$ determines the same monogenic analytic function which $P(z; a)$ does.

This conception of monogenic analytic function is important, because the elements which can be obtained from a given element have properties in common with that element. For instance, if the first element satisfies an algebraic differential equation, every element in the monogenic analytic function which it determines will satisfy the same equation.

CONTINUATION ALONG A CURVE

5. Consider a curve C which joins a point $\underline{a}$ to a point b and an analytic element $P(z; a)$ with center at $\underline{a}$. Suppose that there exist n + 1 points on C

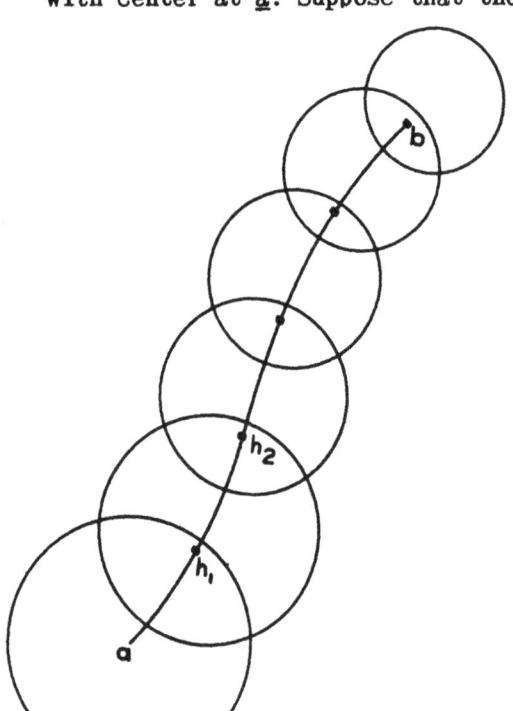

$$h_0 = a, h_1, h_2, \ldots, h_n = b,$$

h_{i+1} coming from a greater value of the parameter than h_i for $i = 0, 1, \ldots, n-1$ and n+1 analytic elements [including $P(z; a) = P(z; h_0)$],

$$P(z; h_0), P(z; h_1), \ldots, P(z; h_n),$$

each, after the first, an immediate continuation of its predecessor, such that for every i that part of C which joins h_i to h_{i+1} lies within the circle of convergence of $P(z; h_i)$.

This chain of elements will be said to *continue* $P(z; a)$ along C.

We are going to prove that *any two chains of elements which continue $P(z; a)$ along C yield the same element at b*. That is, the continuation of $P(z; a)$ with center at b is independent of the particular chain used.

Proof: Consider any point ξ on C and any two chains which continue $P(z; a)$ along C from $\underline{a}$ to ξ. We shall prove that if ξ is very close to $\underline{a}$, the two chains yield the same element at ξ.

Take a circle Γ with center at $\underline{a}$ and of a radius which is one-third the radius of convergence of $P(z; a)$. Let ξ be any point of C within Γ such that the arc of C which joins $\underline{a}$ to ξ lies within Γ. Any immediate continuation $P(z; a')$ of $P(z; a)$, for which a' lies within Γ, will have a circle of convergence which contains the arc of C joining $\underline{a}$ to ξ in its interior. $P(z; a')$ has the same values, for the neighborhood of ξ which $P(z; a)$ does. Any immediate continuation $P(z; a'')$ of $P(z; a')$, for which a'' lies in Γ, will likewise have the same values for the neighborhood of ξ as $P(z; a')$ and hence the same values as $P(z; a)$.

It is thus clear that an analytic element $P(z; \xi)$ obtained from any chain of elements with centers within Γ is merely the immediate continuation of $P(z; a)$ for the point ξ. Certainly then, any two chains yield the same element at ξ.

Now suppose that there are points ξ on C for which two chains yield two distinct elements. Consider those values of the parameter t of the curve which yield such points ξ and let τ be the greatest lower bound of those values. Let d be the point on C which corresponds to τ.

Then for any point ξ on C between a and d, the result of continuation is unique, whereas, either for d or for points arbitrarily close to d, between d and b, two distinct elements can be obtained. Of course, d may coincide with b.

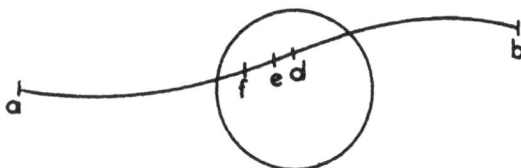

Consider any two chains which terminate with an element having d for center. Suppose that the center which precedes d in the first chain is e, while that which precedes d in the second chain is f. If e and f coincide, the elements corresponding to them in the two chains coincide, because e = f is between a and d. Hence the two chains give the same element for d. Suppose that e and f do not coincide. Let, for instance, e lie between f and d. Then, because the whole arc $\widehat{fd}$ lies inside the circle of convergence of the element with center at f, the element with center at e is an immediate continuation of that with center at f. Hence, the two elements $P(z; f)$, $P(z; e)$ have the same values for the neighborhood of d. Thus, the element at d obtained from either of them is the same.

Hence, the process of continuation gives unique results as far as d is concerned. (If d = b, we are through with the proof.)

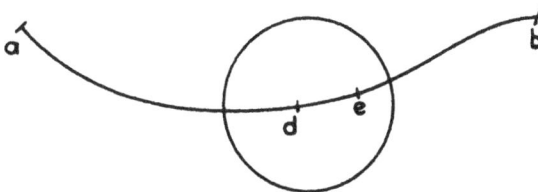

Now, consider any point e between d and b such that the arc $\widehat{de}$ lies within the circle of convergence of $P(z; d)$. Consider any element $P(z; e)$ obtained by continuation from a. We are going to show that $P(z; e)$ is a continuation along $\widehat{de}$ of $P(z; d)$. Consider a chain which continues $P(z; a)$ into $P(z; e)$ and suppose that $P(z; d)$ is not in the chain. Let the chain be

$$P(z; a), \ P(z; k_1), \ \ldots, \ P(z; k_{n-1}), \ P(z; e).$$

Let d be between k_i and k_{i+1}. The entire arc $\widehat{k_i k_{i+1}}$ lies within the circle of convergence of $P(z; k_i)$. Hence, $P(z; d)$ is an immediate continuation of $P(z; k_i)$. Furthermore, the entire arc $\widehat{dk_{i+1}}$ lies within the circle of convergence of $P(z; d)$, because $\widehat{de}$ does. Thus, we can get $P(z; k_{i+1})$ as an immediate continuation of $P(z; d)$.

Hence, $P(z; e)$ is a continuation of $P(z; d)$. It follows, as at the beginning of the proof, that if e is close to d, $P(z; e)$ is unique.

This contradiction of the assumption that two chains can lead to distinct elements at b proves our statement.

SINGULAR POINTS

6. Let $P(z; a)$ be an analytic element with center at a. Let b be any point of the plane, distinct from or coincident with a. Let C be a curve joining a to b.

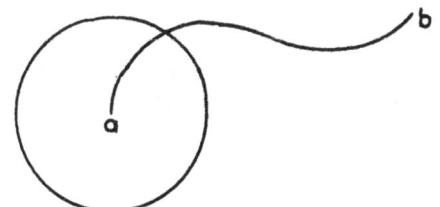

If $P(z; a)$ can be continued along C to b, we call b a *regular point* of the monogenic analytic function defined by $P(z; a)$, *relative to* C.

If $P(z; a)$ can be continued along C to points arbitrarily close to b, but not to b, we call b a *singular point relative to* C.

Example: Let $P(z; a)$ be the expansion of $\frac{1}{z}$ about some point $a \neq 0$. If we take $b = 0$, b is a singular point relative to any curve joining a to b.

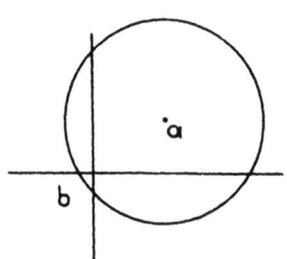

If b is a singular point relative to C, then, as we carry out the continuation of $P(z; a)$ along C towards b, the radii of convergence of the elements obtained must approach zero.

If $P(z; a)$ cannot be continued along C to points arbitrarily close to b, we make no statement relative to b.

Let b be any point of the plane to which $P(z; a)$ can be continued. If, for every such point b, all curves along which $P(z; a)$ can be continued to b yield the same element at b, we shall say that the analytic function determined by $P(z; a)$ is *uniform* or *one-valued*.

If there is a point b such that for two certain curves which join a to b, we secure distinct elements at b, we shall say that the analytic function determined by $P(z; a)$ is *multiform* or *many-valued*.

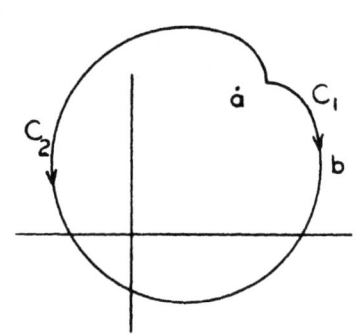

Example: Let $P(z; a)$ be the expansion of one of the two branches of $\sqrt{z}$ at a point $a \neq 0$. If we continue to b along C_1, we get one element of $\sqrt{z}$ at b. If we continue along C_2, we get the negative of the first element.

SINGULAR POINTS ON A CIRCLE OF CONVERGENCE

7. Consider an analytic element $P(z; a)$, with a finite radius of convergence. Let C be the circle of convergence.

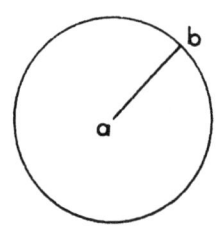

We say that the *analytic function determined by $P(z; a)$ has at least one singular point on C relative to a ray emanating from a*.

Suppose that this is not so. Then, for every b of C, we can get a $P(z; b)$ by continuation from a along the radius ab. Consider the interiors of the circles of convergence of the elements $P(z; b)$. By Borel's theorem, it is possible to find a finite number of them which cover C.

Evidently the points interior to the finite number of circles just obtained, and the points interior to C, constitute an open region, call it A, to which every point of C is interior.

Consider any point k of A. It may be that several of the finite number of elements which we are considering converge at k. Consider any two such elements. If they have the point k in common, they have an area in common which includes part of the interior of C. Since the two elements have the same value at any point interior to C, namely, the value of $P(z; a)$, they have the same value at k. Thus, the finite number of elements $P(z; b)$, and $P(z; a)$, define a function which is analytic in A.

Then we can increase the radius of C slightly, obtaining a larger circle c', in which there exists an analytic function coinciding with $P(z; a)$ within C. This contradicts the fact that C is the circle of convergence of $P(z; a)$ and proves that there is a singular point on C.

If b is a singular point on C and if c is on the radius joining a to b, the circle of convergence of $P(z; c)$ is tangent to that of $P(z; a)$ at b.

NATURAL BOUNDARIES

8. We shall give an example of a power series which cannot be continued outside its circle of convergence. That is every continuation of the power series has a circle of convergence which is tangent internally to the first circle.

Consider the series

$$f(z) = 1 + z + z^{2!} + z^{3!} + \ldots + z^{n!} + \ldots$$

whose radius of convergence is 1.

As we approach the point 1 from the left along the real axis, each term $z^{n!}$ approaches 1, so that the sum of the series approach $+\infty$.

It follows that 1 is a singular point of the series. In short, if any immediate continuation of $f(z)$ had 1 inside its circle of convergence, the continuation would be analytic, therefore bounded in the neighborhood of 1 and $f(z)$ would have to be bounded in the neighborhood of 1.

Now, let ε be any pth root of unity, where p is any positive integer. We have, if $0 \leq r < 1$,

$$f(r\varepsilon) = 1 + r\varepsilon + (r\varepsilon)^{2!} + (r\varepsilon)^{3!} + \ldots + (r\varepsilon)^{n!} + \ldots.$$

When $n > p$, $n!$ is divisible by p, so that

$$(r\varepsilon)^{n!} = r^{n!}.$$

Thus, as r approaches unity through positive values less than 1, the distant terms of our series will all approach $+1$. Consequently $f(z)$ cannot be bounded for the neighborhood of ε, so that ε is a singular point.

Now the points ε are dense all over the circle of convergence. Hence $f(z)$ is not bounded in the neighborhood of any point on the circle of convergence. Thus, every point on the circle of convergence is a singular point. The circle of convergence of any immediate continuation of $f(z)$ is tangent to that of $f(z)$.

When every point on the circle of convergence of the power series is a singular point (radially), we call the circle of convergence a natural boundary of the function.

Bei Fragen zur Produktsicherheit wenden Sie sich bitte an:
If you have any questions regarding product safety,
please contact:

Walter de Gruyter GmbH
Genthiner Straße 13
10785 Berlin
productsafety@degruyterbrill.com